Advances in Information Technology

Advances in Information Technology

Gunjeet Singh

RANDOM

RANDOM PUBLICATIONS

NEW DELHI (INDIA)

Advances in Information Technology

ISBN 978-93-5111-508-3

Published in 2015 in India by

RANDOM PUBLICATIONS

4376-A/4B, Gali Murari Lal, Ansari Road
New Delhi-110 002
Phone : +9111-43580356, 011-23289044, 011-43142548
e-mail: sales@randompublications.com,
info@randompublications.com, randomexports@gmail.com

Reprinted 2026

Type Setting by : Friends Media, Delhi-110089

Preface

In the 1960s and 1970s, the term information technology (IT) was a little known phrase that was used by those who worked in places like banks and hospitals to describe the processes they used to store information. With the paradigm shift to computing technology and "paperless" workplaces, information technology has come to be a household phrase. It defines an industry that uses computers, networking, software programming, and other equipment and processes to store, process, retrieve, transmit, and protect information.

By the early 21st century, nearly every child in the western world, and many in other parts of the world, knew how to use a personal computer. Businesses' information technology departments have gone from using storage tapes created by a single computer operator to interconnected networks of employee workstations that store information in a server farm, often somewhere away from the main business site. Communication has advanced, from physical postal mail, to telephone fax transmissions, to nearly instantaneous digital communication through electronic mail.

Great technological advances have been made since the days when computers were huge pieces of equipment that were stored in big, air conditioned rooms, getting their information from punch cards. The information technology industry has turned out to be a huge employer of people worldwide, as the focus shifts in some nations from manufacturing to service industries. In the current business environment, being proficient in computers is often a necessity for those who want to compete in the workplace.

This book narrates the advances in information technology through a global perspective and emphasizes the importance of making connections among individuals, groups and organizations. It will be ideal for undergraduate business majors with no prerequisite computer courses.

Author

Contents

1

Understanding Information Technology

Information technology (IT) can be defined in various ways, but is broadly considered to encompass the use of computers and telecommunications equipment to store, retrieve, transmit and manipulate data. The term is commonly used as a synonym for computers and computer networks, but it also encompasses other information distribution technologies such as television and telephones.

Humans have been storing, retrieving, manipulating and communicating information since the Sumerians in Mesopotamia developed writing in about 3000 BC, but the term "information technology" in its modern sense first appeared in a 1958 article published in the Harvard Business Review; authors Leavitt and Whisler commented that "the new technology does not yet have a single established name. We shall call it information technology (IT)." Based on the storage and processing technology employed, it is possible to distinguish four distinct phases of IT development: pre-mechanical (3000 BC – 1450 AD), mechanical (1450–1840), electromechanical (1840–1940) and electronic. This article focuses on the latter of those periods, which began in about 1940.

In a business context, the Information Technology Association of America has defined information technology (IT) as "the study, design, development, application, implementation, support or management of computer-based information systems". In an academic context, the Association for Computing Machinery defines it as "undergraduate degree programs that prepare students to meet the computer technology needs of business, government, healthcare, schools, and other kinds of organizations .... IT specialists assume responsibility for selecting hardware and software products appropriate for an organization, integrating those products with organizational needs and infrastructure, and installing, customizing, and maintaining those applications for the organization's computer users. Examples of these responsibilities include the installation of networks; network administration and security; the design of web pages; the development of multimedia resources; the installation of communication components; the oversight of email systems; and the planning and management of the technology lifecycle by which an organization's technology is maintained, upgraded, and replaced."

The Nature of Information

Our lives are awash in information. From the moment we arise each morning, we are surrounded by a seemingly endless array of sources that produce, store, or dispense information. Our alarm clocks are tuned to any one of hundreds of radio broadcast stations that greet us with news, weather, and entertainment. Television offers an equally wide assortment of similar broadcast and cable transmissions. At the breakfast table, for example, we can dine while consulting a single screen for the latest international news, current weather for the U.S., and stock market quotes from Wall Street or the foreign markets. Our telephones—wired or wireless—connect us to any point anywhere on the globe. Millions of conversations like ours flood the channels and airwaves. From our home computers, we can pay our bills and balance our checkbooks, send and receive correspondence, purchase goods and services, and read online magazines and newspapers. Even the appliances in our home keep track of things; in fact, they can be organized to send radio transmissions to exchange information and coordinate their activities. Overhead silent satellites gather data about the weather, population concentrations, and a host of other subjects.

Our own identities are traced by an electronic trail of school transcripts, credit histories, medical records, employment files, and so on. Should these records all disappear in an instant, we would have an unbelievably difficult time persuading others who we are. If you are not convinced of this: Think about the annoyances of merely completing a credit transaction without proper identification.

Business is information and information is business too. At most of our jobs, we spend a significant portion of the workday tending to communications, sifting through reports and records, deciding on what is important and what is not. At the end of the day, we have put in long hours, but very few of us have manufactured goods or products to show for it. What we produce is just as important, though more ephemeral. The financial world of banking is more about the electronic transfer of credits than moving money or gold. Commerce still offers goods and services, but its survival and well-being are no longer measured solely by these material goods and services. Every purchase we make at home or on the road is logged and stored in retail databases. Seasons, cycles, and trends are endlessly plotted and analyzed. Margins of profits and loss are no longer based on inventories, but less tangible things like supply chains. The stock market itself will more likely rise and fall due to reports, indicators, rumors, and predictions than traditional measures like company profits, capital investment, and debt. Everywhere information is the legal tender.

We are living in the Information Age. Our lives, work, and world are all about information. Computers and digital technology have contributed to this flood of information and they are the best means by which we can cope with it. Biologists tell us that one of the hallmarks of life is the ability of an organism to adapt to its surroundings. But, in order to adapt, it must first respond to the conditions of those surroundings. Thus, it is fundamental that all living things are sensitive to external stimuli. When a pin pricks your skin, a sharp pain results. The sudden appearance of a large object in your visual foreground produces the startle response—adrenalin surges in your body and your body prepares to respond to the potential threat. These are some simple examples of this essential characteristic of living things.

In a broader context, though, we can think of sensory responsiveness as a form of communication—that is, the conveyance of information. The sender transmits an encoded message that is interpreted by a receiver. The encoded message contains information that has meaning or significance for the receiver. Thus, organisms have evolved to interpret their natural environment as containing information. Visual perception, for instance, fits the communication model well. Human vision is tuned to a specific bandwidth of electromagnetic energies emanating from the environment—light. But, our receptors do not scan the full range of these frequencies; instead, they sample specific bands within this range (high, middle, and low).

This is sufficient for the visual cortex to compose color information. Nor do we process all of the energies available in the visual environment. Instead, our perceptors are attuned to receive dramatic changes: sharp contrasts in the spatial field, and abrupt changes in the position of objects over time. Perception is not a passive process; it is attuned to a specific medium or channel and is highly selective in what it interprets (receives) as signals (encoded messages).

But the communication model is applicable in even broader contexts. Within the organism, its functioning at various levels involves sending and receiving of encoded messages (information). Cells in a organism must communicate in order to regulate growth and development. Some forms of communication are local; others are long-distance. Consider local cell-to-cell communication first. All cells have a resting potential, that is, they maintain an electrical charge across the plasma membrane of the cell. Some cells, however, such as neurons are excitable. They may be stimulated to create a tiny electrical current. Specifically, certain external conditions can reduce the charge across the membrane causing depolarization. If the voltage reduces beyond its threshold value, an action potential or nerve impulse is generated. The effect of external stimuli is all-or-nothing. As long as the threshold is achieved, a full action potential is generated. This is how neurons transmit signals locally from one cell to the next. Other forms of communication travel longer distances. For example, hormones are secreted by cells that travel in the bloodstream to distant target cells. Insulin is a case in point. It regulates the flow of glucose in the blood by stimulating cells to take up and store glucose as well as inhibiting its production. At a higher level of organization, nerve signals cause muscle fibers to contract or expand. These are just a few examples of how organisms send and receive information at many different levels of organization.

Of course, organisms have evolved to use encoded signals within the species as well. Encoded signals can be interpreting natural signs or direct communication. Migrating birds, for instance, often respond to changes in seasonal weather, the positions of the sun and stars to commence annual long distance treks. Anomalous animal behavior has often been observed with the onset of significant natural events such as earthquakes. Birds, fish, and mammals may be capable of interpreting environmental cues that have eluded us in recognizing such events. Some animals respond to the shortening days, reduced sunlight, and decreasing food supply as signs inducing hibernating behavior. Reversing these cues can bring the same species out of hibernation.

The dances of the European honeybee are often cited as an animal communication system. Scout bees returning to the hive indicate the location and extent of a food source using characteristic dances. Ants use chemical secretions called pheromones for communication. There may be as many as ten different secretions/message complexes in an ant's repertoire. Simple carbon dioxide, for example, promotes clustering for working on larger tasks. Each colony has a unique odor that identifies it and its members. This separates neighbors from intruders. Likewise, foraging ants secrete special pheromones along the trail when returning food to the colony. The density of these trails acts a strong recruiting agent for other ants to follow. High-density trails are more impelling than lower density. As the food supply wanes (i.e., fewer returning foragers), the number of ants recruited naturally dwindles. Birdcalls have been identified as signaling the presence of predators. The calls are different depending upon whether the threat is from below or above.

These and other examples convince us that information is fundamental for life in general. But what exactly is information? We have said a lot of things about it so far without explaining what it is. The problem is that the concept of information is a lot like those of gravity and energy. These concepts are so fundamental, they act as primitives in their domains—everything else is based on them. Information basically refers to knowledge, facts, significance, and meaning—though none of these captures it entirely. Claude Shannon and Warren Weaver in their groundbreaking "A Mathematical Theory of Communication" (1949) described information as what is known beyond chance predictions. Information decreases uncertainty and therefore aids decision-making. In their view, information is measured by the extent to which it is ordered or organized. By contrast, noise is random and disordered. At the other extreme, though, predictable order is monotonous and uninformative. Information lies somewhere between. To some extent information is surprising, unexpected, or new; but it reduces our uncertainty about predicting states of the world.

Information in most instances is useless unless it is transmitted or communicated. As a species, we have developed natural forms of representing and communicating information: signals and natural languages. But, signs and speech are ephemeral; they are bound by the present. Perhaps the defining characteristic of our species is that we have developed technologies to extend these and other powers.

Technologies and Information Technology

Technologies are artificial instruments, processes, or systems that extend our natural capabilities. The wheel, for example, is an extension of the foot as a mode of transportation. Agriculture is an extension of our social institution for gathering food. The microscope and the electric light bulb extend our power of vision into unseen worlds. Technologies also alter and modify our environments. The influences of technologies on society and individuals are sometimes immense and obvious; but these changes can also be subtle and indirect. For example, the invention of the printing press had enormously significant and well-documented effects on the development of the modern world. But, on the other hand, as technologist Marshall McLuhan argued, the invention of the lowly stirrup contributed indirectly to the rise of feudalism in Western Europe.

Of all forms of technology, information technologies are perhaps the most important. From the beginning, humans have extended natural forms of representing and communicating information to incorporate artificial or external forms. Writing, for example, is an artificial form of transmitting and storing the spoken word. As such, writing is a form of information technology. Writing preserves and stores the spoken word externally to the speaker. Thus, it can go beyond the presence of the speaker in both space and time. Indeed, written language can preserve our knowledge and experience beyond our mortality. Certainly, most of what we have achieved as a civilization would not have been possible without our inheritance of the knowledge of generations that have preceded us. This fact alone, perhaps, makes written language the most significant information technology developed by our species. Yet, information technologies can have shortcomings, liabilities, and even ill consequences. The written word, for example, loses some of the richness of meaning conveyed by the spoken word. The written word has saved many but, unfortunately, has also helped condemn many men and women to their deaths unjustly. The bottom line is that technologies can yield both good and bad.

Development of Computers and Data Storage

Devices have been used to aid computation for thousands of years, probably initially in the form of a tally stick. The Antikythera mechanism, dating from about the beginning of the first century BC, is generally considered to be the earliest known mechanical analog computer; it is also the earliest known geared mechanism. Comparable geared devices did not emerge in Europe until the 16th century, and it was not until 1645 that the first mechanical calculator capable of performing the four basic arithmetical operations was developed.

Electronic computers, using either relays or valves, began to appear in the early 1940s. The electromechanical Zuse Z3, completed in 1941, was the world's first programmable computer, and by modern standards one of the first machines that could be considered a complete computing machine. Colossus, developed during the Second World War to decrypt German messages was the first electronic digital computer, but although programmable it was not general-purpose, being designed for a single task. Neither did it store its programs in memory; programming was carried out using plugs and switches to alter the internal wiring. The first recognisably modern electronic digital stored-program computer was the Manchester Small-Scale Experimental Machine (SSEM), which ran its first program on 21 June 1948.

Data Storage

Early electronic computers such as Colossus made use of punched tape, a long strip of paper on which data was represented by a series of holes, a technology now obsolete. Electronic data storage as used in modern computers dates from the Second World War, when a form of delay line memory was developed to remove the clutter from radar signals, the first practical application of which was the mercury delay line. The first random-access digital storage device was the Williams tube, based on a standard cathode ray tube, but the information stored in it and delay line memory was volatile in that it had to be continuously refreshed, and thus was lost once power was removed. The earliest form of non-volatile computer storage was the magnetic drum, invented in 1932 and used in the Ferranti Mark 1, the world's first commercially available general-purpose electronic computer.

Most digital data today is still stored magnetically on devices such as hard disk drives, or optically on media such as CD-ROMs. It has been estimated that the worldwide capacity to store information on electronic devices grew from less than 3 exabytes in 1986 to 295 exabytes in 2007, doubling roughly every 3 years.

Databases

Database management systems emerged in the 1960s to address the problem of storing and retrieving large amounts of data accurately and quickly. One of the earliest such systems was IBM's Information Management System (IMS), which is still widely deployed more than 40 years later. IMS stores data hierarchically, but in the 1970s Ted Codd proposed an alternative relational storage model based on set theory and predicate logic and the familiar concepts of tables, rows and columns. The first commercially available relational database management system (RDBMS) was available from Oracle in 1980. All database management systems consist of a number of components that together allow the data they store to be accessed simultaneously by many users while maintaining its integrity. A characteristic of all databases is that the structure of the data they contain is defined and stored separately from the data itself, in a database schema.

The extensible markup language (XML) has become a popular format for data representation in recent years. Although XML data can be stored in normal file systems, it is commonly held in relational databases to take advantage of their "robust implementation verified by years of both theoretical and practical effort". As an evolution of the Standard Generalized Markup Language (SGML), XML's text-based structure offers the advantage of being both machine and human-readable.

Data Retrieval

The relational database model introduced a programming language independent Structured Query Language (SQL), based on relational algebra.

The terms "data" and "information" are not synonymous. Anything stored is data, but it only becomes information when it is organised and presented meaningfully. Most of the world's digital data is unstructured, and stored in a variety of different physical formats[a] even within a single organisation. Data warehouses began to be developed in the 1980s to integrate these disparate stores. They typically contain data extracted from various sources, including external sources such as the Internet, organised in such a way as to facilitate decision support systems (DSS).

Data Transmission

Data transmission has three aspects: transmission, propagation, and reception. XML has been increasingly employed as a means of data interchange since the early 2000s, particularly for machine-oriented interactions such as those involved in web-oriented protocols such as SOAP, describing "data-in-transit rather than ... data-at-rest". One of the challenges of such usage is converting data from relational databases into XML Document Object Model (DOM) structures.

Data Manipulation

Hilbert and Lopez identify the exponential pace of technological change (a kind of Moore's law): machines' application-specific capacity to compute information per capita roughly doubled every 14 months between 1986 and 2007; the per capita capacity of the world's general-purpose computers doubled every 18 months during the same two decades; the global telecommunication capacity per capita doubled every 34 months; the world's storage capacity per capita required roughly 40 months to double (every 3 years); and per capita broadcast information has doubled every 12.3 years.

Massive amounts of data are stored worldwide every day, but unless it can be analysed and presented effectively it essentially resides in what have been called data tombs: "data archives that are seldom visited". To address that issue, the field of data mining – "the process of discovering interesting patterns and knowledge from large amounts of data" – emerged in the late 1980s.

Digital Information Technology

Electronic digital information technology is the latest generation of information technologies. But, it represents a different brand of information technology. Previously, new information technologies have often competed with and replaced existing ones. The telephone replaced the telegraph for obvious reasons. Television has relegated radio to a subordinate niche. Digital information technology is different because it is a form of technology that extends other technologies. In short, digital information technology has the capability to imitate other technologies. Electronic printed documents mimic conventional typeset ones. Digital audio recordings reproduce sounds like their analog counterparts. To the listener, wireless digital telephony works like normal (wired) telephone service. But its imitation is not mere replication. Digital information technologies offer value-added features. Electronic documents, for example, can be automatically scaled for different media. The same content can be printed on paper, posted on the World Wide Web, transmitted to handheld personal digital assistants (PDAs) and cell phone screens with no extra formatting or fuss. Electronic databases—unlike conventional ones—can be searched and queried automatically revealing facts that would be difficult to find otherwise. Finally, digital information technology can extend the technologies that it imitates by merging them in new and interesting ways. The Web, for example, merges text, numeric data, images, sounds, and video into a seamless medium for posting and sharing content-rich documents.

In support of what we are arguing, consider the rapid acceptance of digital information technologies in so many different arenas and enterprises. As a species, we are a conservative lot. Only a few of us are automatically attracted to new things just because of their novelty. Most of us feel more comfortable with the familiar. On the other hand, we will adapt to and adopt those things that we perceive as us genuinely valuable. Digital information technology must be a case in point. Consider how quickly we have adopted it in so many forms. Would you be willing to give up the Web? E-mail?

Indeed the world of information is going digital. The evolution to digital forms of information has influenced the workplace, the marketplace, our schools, and our homes. In some instances, the changes have been dramatic; in others they have been subtle—almost imperceptible. Considered together, these changes have been highly significant, helping to redefine the ways in which we think about and use information to communicate with each other.

Digital Documents

Typewriters have become relics of an age forgotten in favor of electronic digital word processing. You can now create, edit, and format documents with a flexibility and speed unthought of before computers. The typesetting of books like this one now depend more on digital methods than traditional, conventional techniques. In the past, publishing was the province of the professional; today even school children can produce a typeset book. In short, digital documents are easier to create, store, transmit, and manipulate. One consequence of digital document preparation is that we now create a lot more documents—notes, letters, reports, and books—than ever before.

Digital Numeric Processing

Since their first appearance more than fifty years ago, computers have assisted us in many tasks that involve numerical computation. The earliest computers supported national defense and other government functions. They calculated ballistic tables for artillery, performed computations for developing the first atomic bomb, processed census information, and even predicted presidential elections. Unfortunately, these early computers were expensive and required considerable skill and expertise to operate. Today, the accessibility of computers for numeric processing is widespread. Spreadsheets and statistical and mathematical software offer powerful tools for numerical computation and are easy to use. Most of these tools also have graphing and visualization features that convert large data sets and models to more convenient and accessible forms. Today anyone can command both processing power and productivity for "number crunching" from a desktop computer that far outstrips that of the early giant computers.

Digital Music

Most of the music that you listen to is recorded, stored, and played back using digital methods. Compared to phonograph and audiotape, compact disc digital audio offers much higher fidelity sound. Phonographs and tapes have inherent noise noticeable in playback; their media usually suffer from the wear and tear of constant replay as well. Not so for compact discs. CDs are less noisy and last longer, making digital recording a significantly better archive for important performances. Today we can only imagine how brilliant were the talents of performers such as Enrico Caruso, Toscaninni, and George Gershwin. Their surviving recordings are dim and fading replicas. On the other hand, today's artists can be preserved for all posterity. When you tell your grandchildren how they just don't make music like they used to, you'll be able to prove it by inflicting on them recordings of your favorites that sound just as new as they did the day you bought them.

Digital Photography

When the family is forcibly collected at the photography studio for a portrait, the photographer most likely uses digital methods to capture and process your likenesses. After the shooting, the photographer can use a computer system to display instantly the proofs that used to take days or weeks to process. Thus, you can be photographed and choose which shots you want almost instantly. At the same time, the photographer can show you samples of how these photos may be retouched or processed to make you look even better than you naturally do. Professional photography has made the move to digital methods. Journalistic photography today is often processed digitally; fewer "wet" darkrooms are left in the offices of newspaper and magazine publishers. Recently, consumer versions of digital cameras have become even more affordable. Armed with these and image processing software, you too can produce results that match those of many professionals.

Digital Graphic Arts

Computer graphics (CG) is the artificial generation of images. The field has evolved from computing curiosity to the mainstream. Today, commercial artists employ computer graphics extensively to produce images for newspapers, magazines, and television. Computer-generated animation is used to create or enhance commercials, television shows, and full-length feature films. Indeed, very few films are produced without the benefit of computer-generated effects. The contributions of CG effects to films like the *Harry Potter* sagas and *The Lord of the Rings Trilogy* are abundant and obvious.

Even more importantly, digital processes are making inroads into areas of filmmaking that were the traditional province of analog methods, and not so obvious. For example, the film *O Brother, Where Art Thou?* broke new ground by incorporating digital intermediate processing exclusively. Cinematographer Roger Deakins wanted this Depression-era version of Mark Twain-meets-Homer to have a distinctive look with washed and faded colors. He achieved this by shooting the film using conventional 35mm stock. The film was then scanned and digitized. The digital version was processed to achieve the color and lighting effects that he desired. Afterwards it was converted back to film for theaters. If you see the film, notice that the sky is blue and that flesh tones are natural, yet the backgrounds are faded pastels. This would be impossible using conventional film and lighting effects. Although he did it to achieve special effects, Deakins recognized the general advantages of digital intermediate processing. It offers the cinematographer much greater control over the images of the film. Differences in lighting conditions and film stock can be minimized in postproduction. Likewise, the digital format is natural for archiving and other releases such as DVD. As the technology improves, we can expect its wider acceptance in the film industry. The day of the totally digital "film" is not far away.

Digital Television

Television is migrating to the digital domain as well. Today you can purchase a small, 18-inch diameter satellite dish and receiver for a few hundred dollars. In spite of its size, it can receive hundreds of television channels beamed directly from satellites in fixed orbit around the earth. The picture that you receive is crisp, detailed video, and the audio is multi-channel CD-quality stereo sound. All of this is possible because the transmission and processing are digital. Compared to conventional cable TV service, digital satellite television offers higher fidelity, greater choice, and fewer interruptions of service.

The future of television is undoubtedly digital. The U.S. standard for high-definition TV (HDTV) is a digital one. HDTV offers tremendous gains in resolution that mean unparalleled realism in picture quality. Many of your favorite network programs today are simulcast in HDTV. The FCC has mandated that all broadcast television stations convert to these digital standards for all programming over the next several years.

Virtual Reality

Virtual reality (VR) is a new technology that immerses the user into the illusion of a three-dimensional world built from 3-D graphic models and sophisticated animation techniques. In contrast to a viewer merely watching conventional graphics that are projected onto a video display, the viewer dons special video goggles, headsets, and even gloves for manipulating objects in a specially designed model world. The objective is to create an artificial experience of new phenomena, with the conjured experience having a natural look and feel. Early versions of VR have created artificial rooms and buildings, projected users on journeys across the surface of Mars, and produced trips to virtual art galleries—complete with an extensive collection of art objects. These applications of VR technology have been largely experimental or exploratory. In the near future, as the technology develops, more serious commercial applications will appear. For instance, VR promises to be an invaluable training medium. In medical applications, it can be used to train surgeons. Using ultrasound imaging, VR can also augment the surgeon during live operations as well as by providing additional "visual" cues about the displacement and location of a patient's organs and tissues.

Digital Communications

Person-to-person telephone service is a fact that we all take for granted. Compared to realizing radio and television broadcasting, achieving such service is incredibly complex. A radio transmission can be broadcast across the airwaves. Listeners can receive it merely by purchasing a receiver and tuning in the signal. On the other hand, for conventional telephony, each user must be capable of reaching any other user at any time. The communication must be exclusive and two-way; simple broadcast methods will not do. Consider that there are billions of users worldwide and that sizable distances separate them. Under these circumstances, wired telephone service is amazing. Digital methods have helped to extend service to all points of the globe and have improved performance at the same time. Digitizing voice and data have made transmission faster. Digital techniques are used to combine and transmit thousands of conversations simultaneously on a single carrier. Digital switching systems process and route hundreds of thousands of calls per hour. As commercial communication carriers have converted to digital media and techniques, consumer services have expanded, but the costs have not. In the future, all-digital networks promise to add processing power and intelligence to our telephone service.

Digital methods have also enhanced our wireless communications technologies. Digital signaling methods have extended the range of our devices and permitted greater numbers of users over an already crowded bandwidth. The quality and reliability of the service has also improved in the digital domain.

The Internet

The explosive growth of long-distance networks and the creation of the global Internet are direct consequences of the digitization of commercial communication. Computer networks that were once restricted to a local area can now connect to thousands of other networks around the world. Dedicated network lines called backbones carry millions of individual packets of digital data. These packets hop from one network to another until they reach their intended destination. In this manner, you can send and receive electronic mail, transfer files of data and software, or browse through information on a computer system located thousands of miles away. All of this is possible almost instantly and from the convenience of your own desktop.

The World Wide Web

The use of the Internet has literally exploded over the past several years. A major force in this growth has been the World Wide Web (or Web for short) and the availability of user-friendly programs, called Web browsers (Netscape *Navigator* and Microsoft *Internet Explorer* are two of the most popular), for accessing its wealth of multimedia information. The World Wide Web is a confederation of computer systems that adhere to a common set of guidelines for storing and presenting information to users. These guidelines make it completely straightforward for computer users all over the world to publish multimedia information they have created or collected to all other Web users. Web browser programs exploit the Web's commonality to make all information on the Web accessible in an extremely easy-to-use graphical user interface. This ease of use has made the Web a universal success and phenomenon.

Collections of organized information stored and made available to Web users in a single location are often referred to as Web sites. You will shortly learn about the Web site that accompanies this text. The individual computer files that comprise a Web site are called Web pages. You can always scroll to view all the information on a Web page (although it may extend over several screens). You use the mouse to click over connecting links that access one Web page from another. Most Web sites contain several (perhaps many) Web pages connected to each other for easy access.

These examples illustrate the ways in which digital information has improved or supplanted other conventional forms and how such information can be made available worldwide almost instantly. Of course, none of these capabilities would be possible without the technology of the electronic digital computer. How will this technology ultimately change our world and our lives? Will these changes be for the better or the worse? No one knows the complete answers to such questions. What is clear is that life in the 21st century will be inextricably intertwined with a changing communications paradigm, and the new paradigm will exhibit an increasing dependence on and exploitation of computers and computer networks. Preparation for life in this era will require, even demand, an understanding of the technologies underlying these fundamental changes.

Computing and Information Technology

A computer system is not merely a tool but rather itself a medium for representing, storing, manipulating, and communicating different forms of information: text, numbers, graphics, images, sounds, and video. The common denominator for these differing forms of information is that they all can be digitized for use by our computers. This data can be studied, combined, transformed, and transmitted with an apparent ease that belies the true complexities of these tasks. Digitized data and the systems that handle it constitute the digital domain.

Informational media include text, illustrations, photographs, animation, video, sounds, voice, and music. The modern computer is an all-purpose medium for informational media. Regardless of the media, the computer system represents, stores, and transmits all in its native digital form. That a computer converts text and graphics, for example, to a digital format means that it can process them in similar fashion and at the same time. Multimedia refers to the integration of various forms of information such as text, graphics, sound, and images.

The modern computer system is a multimedia machine; that is, it is capable of integrating two or more conventional forms of informational media in a single electronic document. Because we can express and combine various forms of information using a computer, we can interact, explore, and learn even more from that information. In this way, the computer becomes a vehicle for knowledge rather than just a tool that stores, distributes, and displays information.

The remarkable advances in the price/performance ratio of computer hardware over the past few years, together with a new generation of computer software, are driving dramatic developments in this innovative computer use. In the early days of computing, it was recognized that the speed of computers was especially useful for processing large amounts of numeric and text information. Today, developments in software and hardware are creating opportunities to exploit the computer's capabilities for representing and processing different, richer forms of information that enhance our intellectual abilities. Thus, the traditional model of employing computation for numbers and text is being replaced by a new paradigm. At the heart of these developments is the emergence of two primary technologies: the ability of modern desktop computer systems to collect, store, retrieve, display, and generally manage information in a variety of media and the possibilities for cooperative work using fully interconnected computers and computer networks.

Computer networks are also playing a dominant role in integrating technology into our lives. Networks connect computers in our offices and labs; they also can link us to other computers across the nation and around the world. Using networks, computer systems can share resources and information. That many forms of information can be exchanged instantaneously over long distances has changed the way we work and play. For example, employees in many corporations and other organizations rely more on electronic mail than conventional mail for communication with coworkers.

Indeed, networks have created a new habitat, commonly called cyberspace. These new opportunities have a profound effect on ways we work and interact with one another. Not only is information more readily available, it is also richer and strikingly more dynamic. In cyberspace, you are immersed in and engaged by information rather than merely possessing it.

Communication technologies have existed for a long time. So, what is special about data communications over computer networks? First, the technology driving data communications offers far greater capacity and speed than any other previous form. For example, we now have the capability to store and send whole libraries of information much more quickly than sending a simple message by ordinary postal service. Because this data is represented digitally, we can combine and communicate various forms of information simultaneously, for example, text, audio, and images.

Connectivity means more than simply people communicating with other people. Computer networks also make it possible for individuals to communicate with other computers over long distances. For instance, from your home computer, you can easily borrow both software and processing power from a computer system far away. You can also ask the distant computer system to supply you with the latest stock market quotes, college basketball scores, or international news. When traveling, you can telephone your computer system at home or office to check for electronic mail or messages.

In the past, computers were isolated and largely incompatible. Today, networks support communication from computer to computer as well. Computer systems can request and receive services from other computers automatically and invisibly to the user. This offers a number of advantages. Borrowing processing from a remote system extends the capabilities of your own computer. It also means that computers of different scale and performance can exchange the results of processing almost seamlessly. Distributing the work of processing among cooperating computer systems is still in its infancy. We can expect that it will have a profound effect on computing in the future. Indeed, the idea of individual or autonomously functioning computer systems will very likely become archaic.

REFERENCES

Allen, T., and M. S. Morton, eds. 1994. *Information Technology and the Corporation of the 1990s*. New York: Oxford University Press.

Proctor, K. Scott. 2011. *Optimizing and Assessing Information Technology: Improving Business Project Execution*, John Wiley & Sons.

Reynolds, George. 2009. *Ethics in Information Technology*, Cengage Learning.

Shelly, Gary, Cashman, Thomas, Vermaat, Misty, and Walker, Tim. (1999). *Discovering Computers 2000: Concepts for a Connected World.* Cambridge, Massachusetts: Course Technology.

Webster, Frank, and Robins, Kevin. 1986. *Information Technology—A Luddite Analysis*. Norwood, NJ: Ablex.

2

Computers and Information Processing

A computer is a general purpose device that can be programmed to carry out a finite set of arithmetic or logical operations. Since a sequence of operations can be readily changed, the computer can solve more than one kind of problem.

Conventionally, a computer consists of at least one processing element, typically a central processing unit (CPU) and some form of memory. The processing element carries out arithmetic and logic operations, and a sequencing and control unit that can change the order of operations based on stored information. Peripheral devices allow information to be retrieved from an external source, and the result of operations saved and retrieved.

The first electronic digital computers were developed between 1940 and 1945 in the United Kingdom and United States. Originally they were the size of a large room, consuming as much power as several hundred modern personal computers (PCs). In this era mechanical analog computers were used for military applications.

Modern computers based on integrated circuits are millions to billions of times more capable than the early machines, and occupy a fraction of the space. Simple computers are small enough to fit into mobile devices, and mobile computers can be powered by small batteries. Personal computers in their various forms are icons of the Information Age and are what most people think of as "computers". However, the embedded computers found in many devices from mp3 players to fighter aircraft and from toys to industrial robots are the most numerous.

History of Computing

The first use of the word "computer" was recorded in 1613, referring to a person who carried out calculations, or computations, and the word continued with the same meaning until the middle of the 20th century. From the end of the 19th century the word began to take on its more familiar meaning, a machine that carries out computations.

Limited-function Early Computers

The history of the modern computer begins with two separate technologies, automated calculation and programmability, but no single device can be identified as the earliest computer, partly because of the inconsistent application of that term. A few devices are worth mentioning though, like some mechanical aids to computing, which were very successful and survived for centuries until the advent of the electronic calculator, like the Sumerian abacus, designed around 2500 BC of which a descendant won a speed competition against a modern desk calculating machine in Japan in 1946, the slide rules, invented in the 1620s, which were carried on five Apollo space missions, including to the moon and arguably the astrolabe and the Antikythera mechanism, an ancient astronomical computer built by the Greeks around 80 BC. The Greek mathematician Hero of Alexandria (c. 10–70 AD) built a mechanical theater which performed a play lasting 10 minutes and was operated by a complex system of ropes and drums that might be considered to be a means of deciding which parts of the mechanism performed which actions and when. This is the essence of programmability.

Around the end of the 10th century, the French monk Gerbert d'Aurillac brought back from Spain the drawings of a machine invented by the Moors that answered either Yes or No to the questions it was asked. Again in the 13th century, the monks Albertus Magnus and Roger Bacon built talking androids without any further development (Albertus Magnus complained that he had wasted forty years of his life when Thomas Aquinas, terrified by his machine, destroyed it).

In 1642, the Renaissance saw the invention of the mechanical calculator, a device that could perform all four arithmetic operations without relying on human intelligence. The mechanical calculator was at the root of the development of computers in two separate ways. Initially, it was in trying to develop more powerful and more flexible calculators that the computer was first theorized by Charles Babbage and then developed.

Secondly, development of a low-cost electronic calculator, successor to the mechanical calculator, resulted in the development by Intel of the first commercially available microprocessor integrated circuit.

First General-purpose Computers

In 1801, Joseph Marie Jacquard made an improvement to the textile loom by introducing a series of punched paper cards as a template which allowed his loom to weave intricate patterns automatically. The resulting Jacquard loom was an important step in the development of computers because the use of punched cards to define woven patterns can be viewed as an early, albeit limited, form of programmability.

It was the fusion of automatic calculation with programmability that produced the first recognizable computers. In 1837, Charles Babbage was the first to conceptualize and design a fully programmable mechanical computer, his analytical engine. Limited finances and Babbage's inability to resist tinkering with the design meant that the device was never completed—nevertheless his son, Henry Babbage, completed a simplified version of the analytical engine's computing unit (the mill) in 1888. He gave a successful demonstration of its use in computing tables in 1906. This machine was given to the Science museum in South Kensington in 1910.

In the late 1880s, Herman Hollerith invented the recording of data on a machine-readable medium. Earlier uses of machine-readable media had been for control, not data. "After some initial trials with paper tape, he settled on punched cards ..." To process these punched cards he invented the tabulator, and the keypunch machines. These three inventions were the foundation of the modern information processing industry. Large-scale automated data processing of punched cards was performed for the 1890 United States Census by Hollerith's company, which later became the core of IBM. By the end of the 19th century a number of ideas and technologies, that would later prove useful in the realization of practical computers, had begun to appear: Boolean algebra, the vacuum tube (thermionic valve), punched cards and tape, and the teleprinter.

During the first half of the 20th century, many scientific computing needs were met by increasingly sophisticated analog computers, which used a direct mechanical or electrical model of the problem as a basis for computation. However, these were not programmable and generally lacked the versatility and accuracy of modern digital computers.

Alan Turing is widely regarded as the father of modern computer science. In 1936 Turing provided an influential formalisation of the concept of the algorithm and computation with the Turing machine, providing a blueprint for the electronic digital computer. Of his role in the creation of the modern computer, Time magazine in naming Turing one of the 100 most influential people of the 20th century, states: "The fact remains that everyone who taps at a keyboard, opening a spreadsheet or a word-processing program, is working on an incarnation of a Turing machine".

The Atanasoff–Berry Computer (ABC) was the world's first electronic digital computer, albeit not programmable. Atanasoff is considered to be one of the fathers of the computer. Conceived in 1937 by Iowa State College physics professor John Atanasoff, and built with the assistance of graduate student Clifford Berry, the machine was not programmable, being designed only to solve systems of linear equations. The computer did employ parallel computation. A 1973 court ruling in a patent dispute found that the patent for the 1946 ENIAC computer derived from the Atanasoff–Berry Computer.

The first program-controlled computer was invented by Konrad Zuse, who built the Z3, an electromechanical computing machine, in 1941. The first programmable electronic computer was the Colossus, built in 1943 by Tommy Flowers.

George Stibitz is internationally recognized as a father of the modern digital computer. While working at Bell Labs in November 1937, Stibitz invented and built a relay-based calculator he dubbed the "Model K" (for "kitchen table", on which he had assembled it), which was the first to use binary circuits to perform an arithmetic operation. Later models added greater sophistication including complex arithmetic and programmability.

A succession of steadily more powerful and flexible computing devices were constructed in the 1930s and 1940s, gradually adding the key features that are seen in modern computers. The use of digital electronics (largely invented by Claude Shannon in 1937) and more flexible programmability were vitally important steps, but defining one point along this road as "the first digital electronic computer" is difficult.Shannon 1940 Notable achievements include:

Konrad Zuse's electromechanical "Z machines". The Z3 (1941) was the first working machine featuring binary arithmetic, including floating point arithmetic and a measure of programmability. In 1998 the Z3 was proved to be Turing complete, therefore being the world's first operational computer.

The non-programmable Atanasoff–Berry Computer (commenced in 1937, completed in 1941) which used vacuum tube based computation, binary numbers, and regenerative capacitor memory. The use of regenerative memory allowed it to be much more compact than its peers (being approximately the size of a large desk or workbench), since intermediate results could be stored and then fed back into the same set of computation elements.

The secret British Colossus computers (1943), which had limited programmability but demonstrated that a device using thousands of tubes could be reasonably reliable and electronically reprogrammable. It was used for breaking German wartime codes.

The Harvard Mark I (1944), a large-scale electromechanical computer with limited programmability.

The U.S. Army's Ballistic Research Laboratory ENIAC (1946), which used decimal arithmetic and is sometimes called the first general purpose electronic computer (since Konrad Zuse's Z3 of 1941 used electromagnets instead of electronics). Initially, however, ENIAC had an inflexible architecture which essentially required rewiring to change its programming.

Stored-program Architecture

Several developers of ENIAC, recognizing its flaws, came up with a far more flexible and elegant design, which came to be known as the "stored-program architecture" or von Neumann architecture. This design was first formally described by John von Neumann in the paper First Draft of a Report on the EDVAC, distributed in 1945. A number of projects to develop computers based on the stored-program architecture commenced around this time, the first of which was completed in 1948 at the University of Manchester in England, the Manchester Small-Scale Experimental Machine (SSEM or "Baby"). The Electronic Delay Storage Automatic Calculator (EDSAC), completed a year after the SSEM at Cambridge University, was the first practical, non-experimental implementation of the stored-program design and was put to use immediately for research work at the university. Shortly thereafter, the machine originally described by von Neumann's paper—EDVAC—was completed but did not see full-time use for an additional two years.

Nearly all modern computers implement some form of the stored-program architecture, making it the single trait by which the word "computer" is now defined. While the technologies used in computers have changed dramatically since the first electronic, general-purpose computers of the 1940s, most still use the von Neumann architecture.

Beginning in the 1950s, Soviet scientists Sergei Sobolev and Nikolay Brusentsov conducted research on ternary computers, devices that operated on a base three numbering system of -1, 0, and 1 rather than the conventional binary numbering system upon which most computers are based. They designed the Setun, a functional ternary computer, at Moscow State University. The device was put into limited production in the Soviet Union, but supplanted by the more common binary architecture.

Semiconductors and Microprocessors

Computers using vacuum tubes as their electronic elements were in use throughout the 1950s, but by the 1960s they had been largely replaced by transistor-based machines, which were smaller, faster, cheaper to produce, required less power, and were more reliable. The first transistorised computer was demonstrated at the University of Manchester in 1953. In the 1970s, integrated circuit technology and the subsequent creation of microprocessors, such as the Intel 4004, further decreased size and cost and further increased speed and reliability of computers. By the late 1970s, many products such as video recorders contained dedicated computers called microcontrollers, and they started to appear as a replacement to mechanical controls in domestic appliances such as washing machines. The 1980s witnessed home computers and the now ubiquitous personal computer. With the evolution of the Internet, personal computers are becoming as common as the television and the telephone in the household.

Classification of Computers

The classification of computers depends on computing capacity and data processing speed. Following are some common types of computers.

Mainframe Computers

These are the huge computers you see in a big room where people in white coats mill around. Until the mid-1970s they were the only computers available. With the invention of the mini-computer and later the personal computer, many people said mainframes were too big, too expensive, and not needed anymore. The Internet and the advent of computer networks literally saved the mainframe from oblivion. Mainframes have the necessary power and speed to process millions of transactions from the Internet and networks and have the storage capacity needed for all the data captured and processed by larger Information Systems. America Online, the largest Internet Service Provider in the world, uses many mainframes connected together to process all the transactions across its network and to provide Internet access to the 17 million users who subscribe to the AOL service.

Minicomputers

Mini-computers have less power and capacity than a mainframe but more than a personal computer. They were responsible for bringing computing capacity to a level that smaller companies could afford. Now they are used in smaller networks that don't require the power and speed of mainframes. While they are still useful in many cases, they are sometimes replaced by more powerful workstation and personal computers that are connected to emulate the power of a mini-computer.

Workstation Computers

The workstation computer can be easily confused with a personal computer (PC) because it is configured much thc samc way. IIowcvcr, this computcr has more computing capacity in its CPU than a typical personal computer. Scientists and engineers are the main users of workstations, although people who process graphics (or pictures) find a workstation necessary, since the processors can handle the huge amounts of data associated with graphic files.

Personal computers

While this class of computer used to be called a microcomputer, and sometimes still is, that label gives a false impression of what the machine can do. The prices of PCs have dropped drastically in the last two years, while computing capacity and power have continued to increase. Many small businesses find it cheaper and easier to connect multiple PCs to form a small network than to purchase more expensive equipment. It's estimated that over 60% of the U.S. population has a personal computer in the home.

Supercomputers

Because computing capacity and power were increasing so fast and so much with mainframes and other types of computers, many people didn't see a need for supercomputers. As with the mainframes, the increasing use of networks, with their insatiable appetite for speed and storage, are allowing supercomputers to make a comeback.

A few years ago a computer called "Big Blue," manufactured by IBM, played a series of chess games against Garry Kasparov, the greatest chess player in human history. The interesting part of this match lies in the fact that the computer used a technology called parallel processing. That is, it had thousands of processors working together to calculate each move it would make and the countermove Kasparov would make. Kasparov, on the other hand, was only using one processor—his brain—to calculate all the moves by both sides. Kasparov actually beat the computer for a few rounds but ultimately lost the match. It's fascinating to think that the computer, able to process millions and millions of instructions per second, had to use all of its strengths and massive parallel processing to outwit a person using one single processor.

Components of Computer

A general purpose computer has four main components: the arithmetic logic unit (ALU), the control unit, the memory, and the input and output devices (collectively termed I/O). These parts are interconnected by busses, often made of groups of wires.

Inside each of these parts are thousands to trillions of small electrical circuits which can be turned off or on by means of an electronic switch. Each circuit represents a bit (binary digit) of information so that when the circuit is on it represents a "1", and when off it represents a "0" (in positive logic representation). The circuits are arranged in logic gates so that one or more of the circuits may control the state of one or more of the other circuits.

The control unit, ALU, registers, and basic I/O (and often other hardware closely linked with these) are collectively known as a central processing unit (CPU). Early CPUs were composed of many separate components but since the mid-1970s CPUs have typically been constructed on a single integrated circuit called a microprocessor.

Control Unit

The control unit (often called a control system or central controller) manages the computer's various components; it reads and interprets (decodes) the program instructions, transforming them into a series of control signals which activate other parts of the computer. Control systems in advanced computers may change the order of some instructions so as to improve performance.

A key component common to all CPUs is the program counter, a special memory cell (a register) that keeps track of which location in memory the next instruction is to be read from.

The control system's function is as follows—note that this is a simplified description, and some of these steps may be performed concurrently or in a different order depending on the type of CPU:

- Read the code for the next instruction from the cell indicated by the program counter.
- Decode the numerical code for the instruction into a set of commands or signals for each of the other systems.
- Increment the program counter so it points to the next instruction.
- Read whatever data the instruction requires from cells in memory (or perhaps from an input device). The location of this required data is typically stored within the instruction code.
- Provide the necessary data to an ALU or register.
- If the instruction requires an ALU or specialized hardware to complete, instruct the hardware to perform the requested operation.
- Write the result from the ALU back to a memory location or to a register or perhaps an output device.
- Jump back to step.

Since the program counter is (conceptually) just another set of memory cells, it can be changed by calculations done in the ALU. Adding 100 to the program counter would cause the next instruction to be read from a place 100 locations further down the program. Instructions that modify the program counter are often known as "jumps" and allow for loops (instructions that are repeated by the computer) and often conditional instruction execution (both examples of control flow).

The sequence of operations that the control unit goes through to process an instruction is in itself like a short computer program, and indeed, in some more complex CPU designs, there is another yet smaller computer called a microsequencer, which runs a microcode program that causes all of these events to happen.

Arithmetic Logic Unit (ALU)

The ALU is capable of performing two classes of operations: arithmetic and logic.

The set of arithmetic operations that a particular ALU supports may be limited to addition and subtraction, or might include multiplication, division, trigonometry functions such as sine, cosine, etc., and square roots. Some can only operate on whole numbers (integers) whilst others use floating point to represent real numbers, albeit with limited precision. However, any computer that is capable of performing just the simplest operations can be programmed to break down the more complex operations into simple steps that it can perform. Therefore, any computer can be programmed to perform any arithmetic operation—although it will take more time to do so if its ALU does not directly support the operation. An ALU may also compare numbers and return boolean truth values (true or false) depending on whether one is equal to, greater than or less than the other ("is 64 greater than 65?").

Logic operations involve Boolean logic: AND, OR, XOR and NOT. These can be useful for creating complicated conditional statements and processing boolean logic.

Superscalar computers may contain multiple ALUs, allowing them to process several instructions simultaneously. Graphics processors and computers with SIMD and MIMD features often contain ALUs that can perform arithmetic on vectors and matrices.

Memory

A computer's memory can be viewed as a list of cells into which numbers can be placed or read. Each cell has a numbered "address" and can store a single number. The computer can be instructed to "put the number 123 into the cell numbered 1357" or to "add the number that is in cell 1357 to the number that is in cell 2468 and put the answer into cell 1595". The information stored in memory may represent practically anything. Letters, numbers, even computer instructions can be placed into memory with equal ease. Since the CPU does not differentiate between different types of information, it is the software's responsibility to give significance to what the memory sees as nothing but a series of numbers.

In almost all modern computers, each memory cell is set up to store binary numbers in groups of eight bits (called a byte). Each byte is able to represent 256 different numbers (2^8 = 256); either from 0 to 255 or -128 to +127. To store larger numbers, several consecutive bytes may be used (typically, two, four or eight). When negative numbers are required, they are usually stored in two's complement notation. Other arrangements are possible, but are usually not seen outside of specialized applications or historical contexts. A computer can store any kind of information in memory if it can be represented numerically. Modern computers have billions or even trillions of bytes of memory.

The CPU contains a special set of memory cells called registers that can be read and written to much more rapidly than the main memory area. There are typically between two and one hundred registers depending on the type of CPU. Registers are used for the most frequently needed data items to avoid having to access main memory every time data is needed. As data is constantly being worked on, reducing the need to access main memory (which is often slow compared to the ALU and control units) greatly increases the computer's speed.

Computer main memory comes in two principal varieties: random-access memory or RAM and read-only memory or ROM. RAM can be read and written to anytime the CPU commands it, but ROM is pre-loaded with data and software that never changes, therefore the CPU can only read from it. ROM is typically used to store the computer's initial start-up instructions. In general, the contents of RAM are erased when the power to the computer is turned off, but ROM retains its data indefinitely. In a PC, the ROM contains a specialized program called the BIOS that orchestrates loading the computer's operating system from the hard disk drive into RAM whenever the computer is turned on or reset. In embedded computers, which frequently do not have disk drives, all of the required software may be stored in ROM. Software stored in ROM is often called firmware, because it is notionally more like hardware than software. Flash memory blurs the distinction between ROM and RAM, as it retains its data when turned off but is also rewritable. It is typically much slower than conventional ROM and RAM however, so its use is restricted to applications where high speed is unnecessary.

In more sophisticated computers there may be one or more RAM cache memories, which are slower than registers but faster than main memory. Generally computers with this sort of cache are designed to move frequently needed data into the cache automatically, often without the need for any intervention on the programmer's part.

Input/Output (I/O)

I/O is the means by which a computer exchanges information with the outside world. Devices that provide input or output to the computer are called peripherals. On a typical personal computer, peripherals include input devices like the keyboard and mouse, and output devices such as the display and printer. Hard disk drives, floppy disk drives and optical disc drives serve as both input and output devices. Computer networking is another form of I/O.

I/O devices are often complex computers in their own right, with their own CPU and memory. A graphics processing unit might contain fifty or more tiny computers that perform the calculations necessary to display 3D graphics. Modern desktop computers contain many smaller computers that assist the main CPU in performing I/O.

Multitasking

While a computer may be viewed as running one gigantic program stored in its main memory, in some systems it is necessary to give the appearance of running several programs simultaneously. This is achieved by multitasking i.e. having the computer switch rapidly between running each program in turn.

One means by which this is done is with a special signal called an interrupt, which can periodically cause the computer to stop executing instructions where it was and do something else instead. By remembering where it was executing prior to the interrupt, the computer can return to that task later. If several programs are running "at the same time", then the interrupt generator might be causing several hundred interrupts per second, causing a program switch each time. Since modern computers typically execute instructions several orders of magnitude faster than human perception, it may appear that many programs are running at the same time even though only one is ever executing in any given instant. This method of multitasking is sometimes termed "time-sharing" since each program is allocated a "slice" of time in turn.

Before the era of cheap computers, the principal use for multitasking was to allow many people to share the same computer.

Seemingly, multitasking would cause a computer that is switching between several programs to run more slowly, in direct proportion to the number of programs it is running, but most programs spend much of their time waiting for slow input/output devices to complete their tasks. If a program is waiting for the user to click on the mouse or press a key on the keyboard, then it will not take a "time slice" until the event it is waiting for has occurred. This frees up time for other programs to execute so that many programs may be run simultaneously without unacceptable speed loss.

Multiprocessing

Some computers are designed to distribute their work across several CPUs in a multiprocessing configuration, a technique once employed only in large and powerful machines such as supercomputers, mainframe computers and servers. Multiprocessor and multi-core (multiple CPUs on a single integrated circuit) personal and laptop computers are now widely available, and are being increasingly used in lower-end markets as a result.

Supercomputers in particular often have highly unique architectures that differ significantly from the basic stored-program architecture and from general purpose computers. They often feature thousands of CPUs, customized high-speed interconnects, and specialized computing hardware. Such designs tend to be useful only for specialized tasks due to the large scale of program organization required to successfully utilize most of the available resources at once. Supercomputers usually see usage in large-scale simulation, graphics rendering, and cryptography applications, as well as with other so-called "embarrassingly parallel" tasks.

Information processing

Information processing is the change (processing) of information in any manner detectable by an observer. As such, it is a process that describes everything that happens (changes) in the universe, from the falling of a rock (a change in position) to the printing of a text file from a digital computer system. In the latter case, an information processor is changing the form of presentation of that text file. Information processing may more specifically be defined in terms used by Claude E. Shannon as the conversion of latent information into manifest information. Latent and manifest information is defined through the terms of equivocation (remaining uncertainty, what value the sender has actually chosen), dissipation (uncertainty of the sender what the receiver has actually received), and transformation (saved effort of questioning - equivocation minus dissipation

Information processing may be sequential or parallel, either of which may be centralized or decentralized (distributed). The parallel distributed processing approach of the mid-1980s became popular under the name connectionism. In the early 1950s, Friedrich Hayek was ahead of his time when he posited the idea that spontaneous order in the brain arises out of decentralized networks of simple units (neurons). However, Hayek is rarely cited in the literature of connectionism.

Computer Data Processing

Computer data processing is any process that uses a computer program to enter data and summarise, analyse or otherwise convert data into usable information. It involves recording, analysing, sorting, summarising, calculating, disseminating and storing data. Because data are most useful when well-presented and actually informative, data-processing systems are often referred to as information systems. Nevertheless, the terms are roughly synonymous, performing similar conversions; data-processing systems typically manipulate raw data into information, and likewise information systems typically take raw data as input to produce information as output.

Data processing may or may not be distinguished from data conversion, when the process is merely to convert data to another format, and does not involve any data manipulation.

Uses of Central Processing Unit (CPU)

The CPU is the brains of the computer. Sometimes referred to simply as the processor or central processor, the CPU is where most calculations take place. In terms of computing power, the CPU is the most important element of a computer system. On large machines, CPUs require one or more printed circuit boards. On personal computers and small workstations, the CPU is housed in a single chip called a microprocessor. Two typical components of a CPU are:

— The arithmetic logic unit (ALU), which performs arithmetic and logical operations.

— The control unit, which extracts instructions from memory and decodes and executes them, calling on the ALU when necessary.

Primary Storage

Primary storage, or internal memory, is computer memory that is accessible to the central processing unit of a computer without the use of computer's input/output channels. Primary storage is used to store data that is likely to be in active use. Primary storage is typically very fast, in the case of RAM which is also volatile, losing the stored information in an event of power loss, and quite expensive. ROM is not volatile, but not suited to storage of large quantities of data because it is expensive to produce.

Typically, ROM must also be completely erased before it can be rewritten, making large scale use impractical, if not impossible. Therefore, separate secondary storage, or external memory, is usually required for long-term persistent storage. Confusingly, the term primary storage has recently been used in a few contexts to refer to online storage (hard disks), which is usually classified as secondary storage. Primary storage may include several types of storage, such as main storage, cache memory, and special registers, all of which can be directly accessed by the processor. Primary storage can be accessed randomly, that is, accessing any location in storage at any moment takes the same amount of time. A particular location in storage is selected by its physical memory address. That address remains the same, no matter how the particular value stored there changes.

Semiconductor Memory

Read Only Memory (ROM) is the part of the computer that holds instructions necessary to start your computer. The instructions located in this part of the computer are set when your computer is built, and you can't change them. That's why it's called "Read Only." ROM chips are used in more products than just computers: cars, hand-held calculators, microwave ovens, stereos, alarm clocks, and cellular telephones have them. Most of these chips are either of the PROM or EPROM sub-class of ROM chips. The important characteristic that distinguishes ROM memory from RAM memory is demonstrated by what happens when the computer is turned off or suddenly loses power.

Instructions contained in ROM are non-volatile, which means they aren't lost when the computer is turned off. Instructions contained in RAM are volatile, meaning that they are erased from the primary storage area when the power is turned off. That's why your computer must reload the operating instructions and software programmes whenever you turn the computer back on. If you haven't saved data to the secondary storage area (hard disk drive or floppy disk), you'll lose it from RAM also.

The Arithmetic-Logic Unit (ALU) does one of two things: it performs mathematical calculations or it compares things. The Control Unit does most of the rest of the work by telling other components of the computer what to do and when to do them. It tells the components what to do through a series of instructions that are processed one at time. Even though the instructions are processed one at a time through the machine cycle, they are processed so fast that we can't keep up. Figure 2 shows the machine cycle that every instruction goes through every time it's processed. The machine cycle is very methodical, very precise, and very fast!

Microprocessors

A microprocessor (μP) is a digital electronic component with miniaturised transistors on a single semiconductor integrated circuit (IC). One or more microprocessors typically serve as a central processing unit (CPU) in a computer system or handheld device. Microprocessors made possible the advent of the microcomputer. Before this, electronic CPUs were typically made from bulky discrete switching devices (and later small-scale integrated circuits) containing the equivalent of only a few transistors. By integrating the processor onto one or a very few large-scale integrated circuit packages (containing the equivalent of thousands or millions of discrete transistors), the cost of processor power was greatly reduced.

Since the advent of the IC in the mid-1970s, the microprocessor has become the most prevalent implementation of the CPU, nearly completely replacing all other forms. The evolution of microprocessors has been known to follow Moore's Law when it comes to steadily increasing performance over the years. This law suggests that the complexity of an integrated circuit, with respect to minimum component cost, will double in about 18 months. This rule has been generally followed, unconsciously, since the early 1970s. From their humble beginnings as the drivers for calculators, the continued increase in power has led to the dominance of microprocessors over every other form of computer; every system from the largest mainframes to the smallest handheld computers now uses a microprocessor at its core.

As with many advances in technology, the microprocessor was an idea whose time had come. Three projects arguably delivered a complete microprocessor at about the same time, Intel's 4004, Texas Instruments'

TMS 1000, and Garrett AiResearch's Central Air Data Computer. In 1968, Garrett was invited to produce a digital computer to compete with electromechanical systems then under development for the main flight control computer in the US Navy's new F-14 Tomcat fighter. The design was complete by 1970, and used a MOS-based chipset as the core CPU. The design was smaller and much more reliable than the mechanical systems it competed against, and was used in all of the early Tomcat models. However, the system was considered so advanced that the Navy refused to allow publication of the design, and continued to refuse until 1997. For this reason the CADC, and the MP944 chipset it used, are fairly unknown even today.

TI developed the 4-bit TMS 1000 and stressed pre-programmed embedded applications, introducing a version called the TMS1802NC on September 17, 1971, which implemented a calculator on a chip. The Intel chip was the 4-bit 4004, released on November 15, 1971, developed by Federico Faggin. TI filed for the patent on the microprocessor. Gary Boone was awarded U.S. Patent 3,757,306 for the single-chip microprocessor architecture on September 4, 1973. It may never be known which company actually had the first working microprocessor running on the lab bench. In both 1971 and 1976, Intel and TI entered into broad patent cross-licensing agreements, with Intel paying royalties to TI for the microprocessor patent.

Interestingly, a third party claims to have been awarded a patent which might cover the "microprocessor". See a webpage claiming an inventor predating both TI and Intel, describing a "microcontroller", which may or may not count as a "microprocessor". A computer-on-a-chip is a variation of a microprocessor which combines the microprocessor core (CPU), some memory, and I/O (input/output) lines, all on one chip. The computer-on-a-chip patent, called the "microcomputer patent" at the time, U.S. Patent 4,074,351, was awarded to Gary Boone and Michael J. Cochran of TI. Aside from this patent, the standard meaning of microcomputer is a computer using one or more microprocessors as its CPU(s), while the concept defined in the patent is perhaps more akin to a microcontroller.

8-bit microprocessor

The 4004 was later followed in 1972 by the 8008, the world's first 8-bit microprocessor. These processors are the precursors to the very successful Intel 8080 (1974), Zilog Z80 (1976), and derivative Intel 8-bit processors.

The competing Motorola 6800 was released August 1974. Its architecture was cloned and improved in the MOS Technology 6502 in 1975, rivaling the Z80 in popularity during the 1980s. Both the Z80 and 6502 concentrated on low overall cost, through a combination of small packaging, simple computer bus requirements, and the inclusion of circuitry that would normally have to be provided in a separate chip (for instance, the Z80 included a memory controller). It was these features that allowed the home computer "revolution" to take off in the early 1980s, eventually delivering semi-usable machines that sold for US$99.

The Western Design Centre, Inc. (WDC) introduced the CMOS 65C02 in 1982 and licensed the design to serveral companies which became the core of the Apple IIc and IIe personal computers, medical implantable grade pacemakers and diffribrilators, automotive, industrial and consumer devices. WDC's pioneered the licensing of microprocessor technology which was later followed by ARM and other microprocessor IP providers in the 1990's. Motorola trumped the entire 8-bit world by introducing the MC6809 in 1978, arguably one of the most powerful, orthogonal, and clean 8-bit microprocessor designs ever fielded – and also one of the most complex hardwired logic designs that ever made it into production for any microprocessor.

Microcoding replaced hardwired logic at about this point in time for all designs more powerful than the MC6809 – specifically because the design requirements were getting too complex for hardwired logic. Another early 8-bit microprocessor was the Signetics 2650, which enjoyed a brief flurry of interest due to its innovative and powerful instruction set architecture. A seminal microprocessor in the world of spaceflight was RCA's RCA 1802 (aka CDP1802, RCA COSMAC) (introduced in 1976) which was used in NASA's Voyager and Viking spaceprobes of the 1970s, and onboard the Galileo probe to Jupiter.

The CDP1802 was used because it could be run at very low power, and because its production process (Silicon on Sapphire) ensured much better protection against cosmic radiation and electrostatic discharges than that of any other processor of the era. Thus, the 1802 is said to be the first radiation-hardened microprocessor.

16-bit microprocessor

The first multi-chip 16-bit microprocessor was the National Semiconductor

IMP-16, introduced in early 1973. An 8-bit version of the chipset introduced in 1974 as the IMP-8. In 1975, National introduced the first 16-bit single-chip microprocessor, the PACE, which was later followed by an NMOS version, the INS8900. Other early multi-chip 16-bit microprocessors include one used by Digital Equipment Corporation (DEC) in the LSI-11 OEM board set and the packaged PDP 11/03 minicomputer, and the Fairchild Semiconductor MicroFlame 9440, both of which were introduced in the 1975 to 1976 timeframe.

The first single-chip 16-bit microprocessor was TI's TMS 9900, which was also compatible with their TI 990 line of minicomputers. The 9900 was used in the TI 990/4 minicomputer, the TI-99/4A home computer, and the TM990 line of OEM microcomputer boards. The chip was packaged in a large ceramic 64-pin DIP package package, while most 8-bit microprocessors such as the Intel 8080 used the more common, smaller, and less expensive plastic 40-pin DIP. A follow-on chip, the TMS 9980, was designed to compete with the Intel 8080, had the full TI 990 16-bit instruction set, used a plastic 40-pin package, moved data 8 bits at a time, but could only address 16KB. A third chip, the TMS 9995, was a new design. The family later expanded to include the 99105 and 99110.

The Western Design Centre, Inc. (WDC) introduced the CMOS 65816 16-bit upgrade of the WDC CMOS 65C02 in 1984. The 65816 16-bit microprocessor was the core of the Apple IIgs and later the Super Nintendo Entertainment System, making it one of the most popular 16-bit designs of all time. Intel followed a different path, having no minicomputers to emulate, and instead "upsized" their 8080 design into the 16-bit Intel 8086, the first member of the x86 family which powers most modern PC type computers.

Intel introduced the 8086 as a cost effective way of porting software from the 8080 lines, and succeeded in winning much business on that premise. Following up their 8086 and 8088, Intel released the 80186, 80286 and, in 1985, the 32-bit 80386, cementing their PC market dominance with the processor family's backwards compatibility. The integrated microprocessor memory management unit (MMU) was developed by Childs et al. of Intel, and awarded US patent number 4,442,484.

32-bit microprocessor

16-bit designs were in the market only briefly when full 32-bit implementations started to appear. The world's first single-chip 32-bit microprocessor was the AT&T Bell Labs BELLMAC-32A, with first samples in 1980, and general production in 1982. After the divestiture of AT&T in 1984, it was renamed the WE 32000 (WE for Western Electric), and had two follow-on generations, the WE 32100 and WE 32200. These microprocessors were used in the AT&T 3B5 and 3B15 minicomputers; in the 3B2, the world's first desktop supermicrocomputer; in the "Companion", the world's first 32-bit laptop computer; and in "Alexander", the world's first book-sized supermicrocomputer, featuring ROM-pack memory cartridges similar to today's gaming consoles.

All these systems ran the original Bell Labs Unix Operating System, which included the first Windows-type software called xt-layers. The most famous of the 32-bit designs is the MC68000, introduced in 1979. The 68K, as it was widely known, had 32-bit registers but used 16-bit internal data paths, and a 16-bit external data bus to reduce pin count. Motorola generally described it as a 16-bit processor, though it clearly has 32-bit architecture. The combination of high speed, large (16 megabyte) memory space and fairly low costs made it the most popular CPU design of its class.

The Apple Lisa and Macintosh designs made use of the 68000, as did a host of other designs in the mid-1980s, including the Atari ST and Commodore Amiga. Intel's first 32-bit microprocessor was the iAPX 432, which was introduced in 1981 but was not a commercial success. It had an advanced capability-based object-oriented architecture, but poor performance compared to other competing architectures such as the Motorola 68000. Motorola's success with the 68000 led to the MC68010, which added virtual memory support. The MC68020, introduced in 1985 added full 32-bit data and address busses.

The 68020 became hugely popular in the Unix supermicrocomputer market, and many small companies (e.g., Altos, Charles River Data Systems) produced desktop-size systems. Following this with the MC68030, which added the MMU into the chip, the 68K family became the processor for everything that wasn't running DOS. The continued success led to the MC68040, which included an FPU for better math performance. A 68050 failed to achieve its performance goals and was not released, and the follow-up MC68060 was released into a market saturated by much faster RISC designs. The 68K family faded from the desktop in the early 1990s. Other large companies designed the 68020 and follow-ons into embedded equipment.

At one point, there were more 68020s in embedded equipment than there were Intel Pentiums in PCs. The ColdFire processor cores are derivatives of the venerable 68020. During this time (early to mid 1980s), National Semiconductor introduced a very similar 16-bit pinout, 32-bit internal microprocessor called the NS 16032 (later renamed 32016), the full 32-bit version named the NS 32032, and a line of 32-bit industrial OEM microcomputers. By the mid-1980s, Sequent introduced the first symmetric multiprocessor (SMP) server-class computer using the NS 32032. This was one of the design's few wins, and it disappeared in the late 1980s.

Other designs included the interesting Zilog Z8000, which arrived too late to market to stand a chance and disappeared quickly. In the late 1980s, "microprocessor wars" started killing off some of the microprocessors. Apparently, with only one major design win, Sequent, the NS 32032 just faded out of existence, and Sequent switched to Intel microprocessors.

64-bit microprocessor

While 64-bit microprocessor designs have been in use in several markets since the early 1990s, the early 2000s have seen the introduction of 64-bit microchips targeted at the PC market. With AMD's introduction of the first 64-bit IA-32 backwards-compatible architecture, AMD64, in September 2003, followed by Intel's own x86-64 chips, the 64-bit desktop era began. Both processors can run 32-bit legacy apps as well as the new 64-bit software. With 64-bit Windows XP and Linux that run 64-bit native, the software too is geared to utilise the full power of such processors. In reality the move to 64-bits is more than just an increase in register size from the ia32 as it also includes a small increase in register quantity for the aging CISC designs. The move to 64 bits by PowerPC processors had been intended since the processors design in the early 90s and was not a major cause of incompatibility. Existing integer registers are extended as are all related data pathways but in common with the IA32 designs both floating point and vector units had been operating at or above 64 bits for several years. Unlike the IA32 no new general purpose registers are added so any performance gained when using the 64-bit mode is minimal.

RISC microprocessor

In the mid-1980s to early-1990s, a crop of new high-performance RISC (reduced instruction set computer) microprocessors appeared, which were initially used in special purpose machines and Unix workstations, but have since become almost universal in all roles except the Intel-standard desktop. The first commercial design was released by MIPS Technologies, the 32-bit R2000 (the R1000 was not released). The R3000 made the design truly practical, and the R4000 introduced the world's first 64-bit design. Competing projects would result in the IBM POWER and Sun SPARC systems, respectively.

Soon every major vendor was releasing a RISC design, including the AT&T CRISP, AMD 29000, Intel i860 and Intel i960, Motorola 88000, DEC Alpha and the HP-PA. Market forces have "weeded out" many of these designs, leaving the POWER and the derived PowerPC as the main desktop RISC processor, with the SPARC being used in Sun designs only. MIPS continues to supply some SGI systems, but is primarily used as an embedded design, notably in Cisco routers. The rest of the original crop of designs have either disappeared, or are about to. Other companies have attacked niches in the market, notably ARM, originally intended for home computer use but since focussed at the embedded processor market. Today RISC designs based on the MIPS, ARM or PowerPC core power the vast majority of computing devices. In 64-bit computing, DEC Alpha, AMD64, MIPS, SPARC, PowerPC/IBM POWER, and HP-Intel Itanium are all popular designs.

Special-purpose microprocessors

Though the term "microprocessor" has traditionally referred to a single- or multi-chip CPU or SoC, several types of specialised processing devices have followed from the technology. The most common examples are microcontrollers, DSPs and GPUs. Many examples of these are either not programmable, or have limited programming facilities. For example, in general GPUs through the 1990s were mostly non-programmable and have only recently gained limited facilities like programmable vertex shaders. There is no universal consensus on what defines a "microprocessor", but it is usually safe to assume that the term refers to a general-purpose CPU of some sort and not a special-purpose processor unless specifically noted. The RCA 1802 had what is called a static design, meaning that the clock frequency could be made arbitrarily low, even to 0Hz, a total stop condition. This let the Voyager/Viking/Galileo spacecraft use minimum electric power for long uneventful stretches of a voyage. Timers and/or sensors would awaken/speed up the processor in time for important tasks, such as navigation updates, attitude control, data acquisition, and radio communication.

Client/Server Computing

Client/Server is a network architecture which separates the client (often a graphical user interface) from the server. Each instance of the client software can send requests to a server or application server. Although this idea is applied in a variety of ways, on many different kinds of applications, the easiest example to visualise is the current use of web pages on the internet. A generic Client/Server architecture has two types of nodes on the network: clients and servers. As a result, these generic architectures are sometimes referred to as "two-tier" architectures.

A Client/Server architecture is intended to provide a scalable architecture, whereby each computer or process on the network is either a client or a server. Server software generally, but not always, runs on powerful computers dedicated for exclusive use to running the business application. Client software on the other hand generally runs on common PCs or workstations. Clients get all or most of their information and rely on the application server for things such as configuration files, stock quotes, business application programs, or to offload computer-intensive application tasks back to the server in order to keep the client computer (and client computer user) free to perform other tasks. Properties of a server are:

— Passive (Slave)

— Waiting for requests

— On requests serves them and send a reply

Properties of a client are:

— Active (Master)

— Sending requests

— Waits until reply arrives

Servers can be stateless or stateful. A stateless server does not keep any information between requests. Example: An HTTP server for static HTML pages. A stateful server can remember information between requests. The scope of this information can be global or session. Example: Apache Tomcat. The interaction between client and server is often described using sequence diagrams. Sequence diagrams are standardised in the UML. Another type of network architecture is known as a peer-to-peer architecture because each node or instance of the program is both a "client" and a "server" and each has equivalent responsibilities. Both architectures are in wide use.

Some networks will consist of three different kinds of nodes, clients, application servers which process data for the clients, and database servers which store data for the application servers. This is called a three-tier architecture. In general, an n-tier or multi-tier architecture may deploy any number of distinct services, including transitive relations between application servers implementing different functions of business logic, each of which may or may not employ a distinct or shared database system.

The advantage of an n-tier architecture compared with a two-tier architecture (or a three-tier with a two-tier) is that it separates out the processing that occurs to better balance the load on the different servers; it is more scalable.

The disadvantages of n-tier architectures are:

— It puts a greater load on the network.

— It is much more difficult to program and test software than in two-tier architecture because more devices have to communicate to complete a user's transaction.

Methods of Addressing

Methods of addressing in client server environments can be described as follows:

— Machine process addressing; where the address is divided up as follows process@machine. Therefore 56@453 would indicate the process 56 on computer 453

— Name Server; Name servers have an index of all names and addresses of servers in the relevant domain.

— Localisation packets; Broadcast messages are sent out to all computers in the distributed system to determine the address of the destination computer

— Trader; A trader is a system that indexes all the services available in a distributed system. A computer requiring a particular service will check with the trading service for the address of a computer providing such a service.

Network Computers and Total Cost of Ownership

The cost issue is becoming more important to businesses and companies as computer technology and networks grow. Depending on the configuration of the network, a company can save or lose many dollars. What's most important to remember is that the Total Cost of Ownership (TCO) should extend past the hard dollars spent on hardware and software. The cost should incorporate such as items as employee training, their ability to perform necessary functions given the network configuration, and lost productivity when the network is down. The TCO should also include the amount of money spent on communications wiring (telephone wires, fiber optic cable, etc) and security and access issues.

Secondary Storage

Secondary storage includes the hard disk drives in almost every computer, floppy disks, and magnetic tape storage. The kind of storage necessary is dictated by the size of the computer and how it's used. As with most other components, the cost is dropping significantly while the technology is improving all the time.

Magnetic disk and tape

The most common form of secondary storage is the magnetic disk. Magnetic tape is still used on larger mainframe computers for data that don't require fast access, such as historical records maintained by credit card companies. If you want fast, direct access to data, use the magnetic disks that are most common in personal computers. Just about every personal computer has a hard disk drive located in the box that sits beside the monitor. All the operating system software and application program software resides on the hard disk drive for long-term storage.

Data files are also saved to the hard disk (or floppy disk). When you start your personal computer, the operating system instructions most commonly used are moved from the hard drive to the RAM memory. Whenever you open an application program, the most frequently used instructions are also moved to the RAM memory for quick retrieval. The instructions that aren't currently being used remain on the hard drive. The common 3½" floppy disk that we are used to seeing is becoming outdated because of size limitations.

Most of these disks can only hold 1.44 million bytes of data. Zip disks can hold anywhere from 100 million to 250 million bytes. They operate basically the same way a floppy disk does, but you need a special zip disk drive in order to use them. When you create or use files with lots of graphics, you'll need the larger type of disk in order to hold the data.

Optical disks

We've probably all seen a compact disk (CD-ROM) that holds music. This same type of disk is becoming more common with personal computers because of the amount of data it can hold. Many software programs are on compact disks (CD). Otherwise, you'd have to use 20-30 floppy disks to install the software. The drawback to this type of disk is that it can usually be used only once because the data are stored on the disk by literally burning pits into its surface. Machines that allow the average user to record CDs (compact disk-recordable) are starting to show up on store shelves, but they are still rather pricey.

The WORM (Write Once/Read Many) technology gives the user the technique of writing to a CD once but reading it many times. And some machines now allow you to rewrite a CD (CDReWritable) by using a combination of magnetic and optical writing techniques. Digital video disks (DVD) can be used on either computers or with televisions. You can rent DVDs from your local video store and watch a movie on either device. It's safe to say that the 3 ½" floppy disks are on their way to the museum, but because of convenience and sheer numbers, it will be a while before they are gone altogether.

Input Devices

Data go in and information comes out of computer systems via peripheral devices. The speed, capacity, and ease of use of these devices is constantly improving.

You're pretty familiar with the keyboard and mouse input devices. Keyboards and mouse pointers are becoming more comfortable to use through ergonomically improved designs. However, many people find these input devices cumbersome and slow. They are turning to pen-based input, which uses a tablet device and a pen stylus. Companies such as Federal Express and UPS use the technology for customers to accept packages. The data are then transferred directly to a mainframe computer which allows packages to be tracked from the sender to the recipient. The latest trend for input devices is voice recognition. This technology allows you to put data into a computer using a combination of a microphone headset and software on your computer that recognises your voice.

You speak into the microphone and the software interprets your voice and records your words. You can also use this technology to execute commands on your computer, such as "Save this file" or "Print this document." The software associated with voice recognition technology is very complex because no two people speak alike. We all have accents, voice inflections, slang words, or even headcolds that change the tone of our voice. When you initially use voice recognition software, you have to "train" the computer to recognise your voice and all of its characteristics. It typically takes 2 to 3 hours for the initial training with additional training as you continually use the software. In essence, you build your own dictionary of words that are unique to you. This is one of the greatest inventions for people with physical disabilities, since they can rely on voice to control the computer. Executives find this technology especially useful. It is also great for people who do a lot of typing and could possibly or do suffer from repetitive stress injuries. In fact, these computer files were generated using voice recognition software to reduce the amount of typing necessary to create them! Unfortunately voice recognition doesn't work very well in some corporate settings. It can be very distracting to the workers around you if you are constantly talking into a microphone.

The background noise of most office settings can also interfere with the voice input. Touch screens are also becoming more common for general usage. Touch screens have been used in ATM banking machines for years. Now they are coming to a screen near you—computer or television or household appliances. Source data automation is popular with businesses because the technology reduces the chance of input error. A human may type in 2345 instead of 2354Ña simple error that could have significant consequences.

Optical character recognition software or scanner software will always read the number correctly. Digital scanner technology extends far beyond the page scanners attached to personal computers or used in offices. The scanners grocery stores use at the checkout counter are a great example of this technology. The vast number of hand-held scanners used by other retailers are also typical of the ability of technology to increase human productivity. The biggest benefit of this technology lies in the fact that input errors are greatly reduced. The computer reads what is there and processes it. If an error occurs, which does happen in groceries about 5 % of the time, it isn't because of the computer. It's because the human programming the computer made a mistake. Banks have used magnetic ink character recognition (MICR) technology for years to process checks. If you still receive cancelled checks from your bank, take a look at the bottom right hand corner of the check. You'll notice that the amount of the check and the date it was processed is imprinted in block style numbers. The computer added this information to ensure that your account was properly credited.

Batch and on-line processing

Whether a business uses batch processing or on-line processing depends:

— On how soon the data must be input

— On how fast the updates must be made

— On how accessible the data must be

If the data are not needed immediately, batch processing is usually the chosen method. If the data must be immediately available, the required method is on-line processing. Note the different uses of the transaction file and the master file. Many companies are switching to on-line processing even for data that have traditionally been batch processed because they don't want to maintain two different processing methods.

Output Devices

What goes into a computer eventually needs to come out of the computer. Typically we think of printers as the main form of output. But more and more often, the output of a computer merely extends to the monitor screen. A good example of this is the Web. We surf the Web looking for information or entertainment. We read the information, process it through our "personal processor" (commonly called a brain), and then we're on to another Web site. Or, we save it to our disk drive to access at a later time.

The quality of the cathode ray tube (CRT) is determined by the number of pixels on the screen. The higher the number of pixels, the better resolution you'll have. Bit mapping is the technique the computer uses to manipulate the pixels on your screen. In addition to being an input device, voice recognition is also an output device. If you call a major corporation on the telephone, it would be unusual for you to actually talk to a human being.

More often, the company uses voice recognition technology to guide you through menus of services or answer your questions.

The cost of laser and ink-jet printer devices (nonimpact printers) has dropped significantly in the last few years so they are affordable for home use. You should be aware that the cost of using these devices extends past the price of the device itself. When considering the cost of printing you should include the cartridges and paper. The cartridges usually cost $25 to $30 each, while the paper can be twice the còst of regular bond paper. Impact printers, such as the dot matrix, are much cheaper to own and operate. The paper is less costly than bond paper and the printer ribbons are approximately $5 each. However, the quality is significantly lower.

References

Alavudeen, A.; Venkateshwaran, N. 2010, *Computer Integrated Manufacturing*, PHI Learning.

Bynum, Terrell Ward 2008, "Norbert Wiener and the Rise of Information Ethics", in van den Hoven, Jeroen; Weckert, John,*Information Technology and Moral Philosophy*, Cambridge University Press.

Chaudhuri, P. Pal. 2004, *Computer Organization and Design*, PHI Learning.

Childress, David Hatcher. 2000. *Technology of the Gods: The Incredible Sciences of the Ancients*, Adventures Unlimited Press.

Dyché, Jill. 2000, *Turning Data Into Information With Data Warehousing*, Addison Wesley.

3

Operating Systems

An operating system (OS) is a collection of software that manages computer hardware resources and provides common services for computer programs. The operating system is a vital component of the system software in a computer system. Application programs usually require an operating system to function.Time-sharing operating systems schedule tasks for efficient use of the system and may also include accounting for cost allocation of processor time, mass storage, printing, and other resources.

For hardware functions such as input and output and memory allocation, the operating system acts as an intermediary between programs and the computer hardware, although the application code is usually executed directly by the hardware and will frequently make a system call to an OS function or be interrupted by it. Operating systems can be found on almost any device that contains a computer—from cellular phones and video game consoles to supercomputers and web servers. Examples of popular modern operating systems include Android, BSD, iOS, Linux, Mac OS X, Microsoft Windows, Windows Phone, and IBM z/OS. All these, except Windows and z/OS, share roots in UNIX.

Development of Operating Systems

Early computers were built to perform a series of single tasks, like a calculator. Operating systems did not exist in their modern and more complex forms until the early 1960s. Basic operating system features were developed in the 1950s, such as resident monitor functions that could automatically run different programs in succession to speed up processing. Hardware features were added that enabled use of runtime libraries, interrupts, and parallel processing. When personal computers became popular in the 1980s, operating system were made for them similar in concept to those used on larger computers.

In the 1940s, the earliest electronic digital systems had no operating systems. Electronic systems of this time were programmed on rows of mechanical switches or by jumper wires on plug boards. These were special-purpose systems that, for example, generated ballistics tables for the military or controlled the printing of payroll checks from data on punched paper cards. After programmable general purpose computers were invented, machine languages (consisting of strings of the binary digits 0 and 1 on punched paper tape) were introduced that sped up the programming process.

In the early 1950s, a computer could execute only one program at a time. Each user had sole use of the computer for a limited period of time and would arrive at a scheduled time with program and data on punched paper cards and/or punched tape. The program would be loaded into the machine, and the machine would be set to work until the program completed or crashed. Programs could generally be debugged via a front panel using toggle switches and panel lights. It is said that Alan Turing was a master of this on the early Manchester Mark 1 machine, and he was already deriving the primitive conception of an operating system from the principles of the Universal Turing machine.

Later machines came with libraries of programs, which would be linked to a user's program to assist in operations such as input and output and generating computer code from human-readable symbolic code. This was the genesis of the modern-day computer system. However, machines still ran a single job at a time. At Cambridge University in England the job queue was at one time a washing line from which tapes were hung with different colored clothes-pegs to indicate job-priority.

Mainframes

Through the 1950s, many major features were pioneered in the field of operating systems, including batch processing, input/output interrupt, buffering, multitasking, spooling, runtime libraries, link-loading, and programs for sorting records in files. These features were included or not included in application software at the option of application programmers, rather than in a separate operating system used by all applications. In 1959 the SHARE Operating System was released as an integrated utility for the IBM 704, and later in the 709 and 7090 mainframes, although it was quickly supplanted by IBSYS/IBJOB on the 709, 7090 and 7094.

During the 1960s, IBM's OS/360 introduced the concept of a single OS spanning an entire product line, which was crucial for the success of the System/360 machines. IBM's current mainframe operating systems are distant descendants of this original system and applications written for OS/360 can still be run on modern machines.

OS/360 also pioneered the concept that the operating system keeps track of all of the system resources that are used, including program and data space allocation in main memory and file space in secondary storage, and file locking during update. When the process is terminated for any reason, all of these resources are re-claimed by the operating system.

The alternative CP-67 system for the S/360-67 started a whole line of IBM operating systems focused on the concept of virtual machines. Other operating systems used on IBM S/360 series mainframes included systems developed by IBM: COS/360 (Compatibility Operating System), DOS/360 (Disk Operating System), TSS/360 (Time Sharing System), TOS/360 (Tape Operating System), BOS/360 (Basic Operating System), and ACP (Airline Control Program), as well as a few non-IBM systems: MTS (Michigan Terminal System), MUSIC (Multi-User System for Interactive Computing), and ORVYL (Stanford Timesharing System).

Control Data Corporation developed the SCOPE operating system in the 1960s, for batch processing. In cooperation with the University of Minnesota, the Kronos and later the NOS operating systems were developed during the 1970s, which supported simultaneous batch and timesharing use. Like many commercial timesharing systems, its interface was an extension of the Dartmouth BASIC operating systems, one of the pioneering efforts in timesharing and programming languages. In the late 1970s, Control Data and the University of Illinois developed the PLATO operating system, which used plasma panel displays and long-distance time sharing networks. Plato was remarkably innovative for its time, featuring real-time chat, and multi-user graphical games. Burroughs Corporation introduced the B5000 in 1961 with the MCP, (Master Control Program) operating system. The B5000 was a stack machine designed to exclusively support high-level languages with no machine language or assembler, and indeed the MCP was the first OS to be written exclusively in a high-level language – ESPOL, a dialect of ALGOL. MCP also introduced many other ground-breaking innovations, such as being the first commercial implementation of virtual memory. During development of the AS400, IBM made an approach to Burroughs to licence MCP to run on the AS400 hardware. This proposal was declined by Burroughs management to protect its existing hardware production. MCP is still in use today in the Unisys ClearPath/MCP line of computers.

UNIVAC, the first commercial computer manufacturer, produced a series of EXEC operating systems. Like all early main-frame systems, this was a batch-oriented system that managed magnetic drums, disks, card readers and line printers. In the 1970s, UNIVAC produced the Real-Time Basic (RTB) system to support large-scale time sharing, also patterned after the Dartmouth BC system.

General Electric and MIT developed General Electric Comprehensive Operating Supervisor (GECOS), which introduced the concept of ringed security privilege levels. After acquisition by Honeywell it was renamed to General Comprehensive Operating System (GCOS).

Digital Equipment Corporation developed many operating systems for its various computer lines, including TOPS-10 and TOPS-20 time sharing systems for the 36-bit PDP-10 class systems. Prior to the widespread use of UNIX, TOPS-10 was a particularly popular system in universities, and in the early ARPANET community.

In the late 1960s through the late 1970s, several hardware capabilities evolved that allowed similar or ported software to run on more than one system. Early systems had utilized microprogramming to implement features on their systems in order to permit different underlying computer architectures to appear to be the same as others in a series. In fact most 360s after the 360/40 (except the 360/165 and 360/168) were microprogrammed implementations. But soon other means of achieving application compatibility were proven to be more significant.

The enormous investment in software for these systems made since 1960s caused most of the original computer manufacturers to continue to develop compatible operating systems along with the hardware. The notable supported mainframe operating systems include:

- Burroughs MCP – B5000, 1961 to Unisys Clearpath/MCP, present.

- IBM OS/360 – IBM System/360, 1966 to IBM z/OS, present.
- IBM CP-67 – IBM System/360, 1967 to IBM z/VM, present.
- UNIVAC EXEC 8 – UNIVAC 1108, 1967, to OS 2200 Unisys Clearpath Dorado, present.

Microcomputers

The first microcomputers did not have the capacity or need for the elaborate operating systems that had been developed for mainframes and minis; minimalistic operating systems were developed, often loaded from ROM and known as monitors. One notable early disk operating system was CP/M, which was supported on many early microcomputers and was closely imitated by Microsoft's MS-DOS, which became wildly popular as the operating system chosen for the IBM PC (IBM's version of it was called IBM DOS or PC DOS). In the '80s, Apple Computer Inc. (now Apple Inc.) abandoned its popular Apple II series of microcomputers to introduce the Apple Macintosh computer with an innovative Graphical User Interface (GUI) to the Mac OS.

The introduction of the Intel 80386 CPU chip with 32-bit architecture and paging capabilities, provided personal computers with the ability to run multitasking operating systems like those of earlier minicomputers and mainframes. Microsoft responded to this progress by hiring Dave Cutler, who had developed the VMS operating system for Digital Equipment Corporation. He would lead the development of the Windows NT operating system, which continues to serve as the basis for Microsoft's operating systems line. Steve Jobs, a co-founder of Apple Inc., started NeXT Computer Inc., which developed the NEXTSTEP operating system. NEXTSTEP would later be acquired by Apple Inc. and used, along with code from FreeBSD as the core of Mac OS X.

The GNU Project was started by activist and programmer Richard Stallman with the goal of creating a complete free software replacement to the proprietary UNIX operating system. While the project was highly successful in duplicating the functionality of various parts of UNIX, development of the GNU Hurd kernel proved to be unproductive. In 1991, Finnish computer science student Linus Torvalds, with cooperation from volunteers collaborating over the Internet, released the first version of the Linux kernel. It was soon merged with the GNU user space components and system software to form a complete operating system. Since then, the combination of the two major components has usually been referred to as simply "Linux" by the software industry, a naming convention that Stallman and the Free Software Foundation remain opposed to, preferring the name GNU/Linux. The Berkeley Software Distribution, known as BSD, is the UNIX derivative distributed by the University of California, Berkeley, starting in the 1970s. Freely distributed and ported to many minicomputers, it eventually also gained a following for use on PCs, mainly as FreeBSD, NetBSD and OpenBSD.

Goal of an Operating System

The fundamental goal of computer system is to solve user problems. Accordingly to achieve this goal has been designed. Since hardware alone cannot be used to solve the user problems software's are developed. These programs required certain common operations. The common operations for controlling and allocating resources are then brought together into one piece of software i.e. operating system. An operating system may process its tasks sequentially or concurrently. It means that the resources of the computer system may be dedicated to a single program until its completion or they may be allocated among several programs in different stages of execution.

The feature of operating system is to execute multiple programs in interleaved fashion or different time cycle is called multiple programming systems. Some of the important reasons why do we need an Operating System are as follows:

- User interacts with the computer through operating system in order to accomplish his/her task since it is his primary interface with a computer.
- It helps the user in understand the inner functions of a computer very closely.
- Many concepts and techniques found in operating system have general applicability in other applications. An operating system is an essential component of a computer system. The primary objectives of an operating system are to make computer system convenient to use and utilises computer hardware in an efficient manner.

An operating system is a large collection of software which manages resources of the computer system, such as memory, processor, rite system and input/output devices. It keeps track of the status of each resource and decides who will have a control over computer resources, for how long and when. The positioning of operating system in overall computer system is shown in figure 1. From the diagram, it is clear that operating system directly controls computer hardware resources. Other programs rely on facilities provided by the operating system to gain access to computer system resources. There are two ways one can interact with operating system:

— By means of Operating System Call in a program

— Directly by means of Operating System Commands

System Call

System calls provide the interface to a running program and the operating system. User program receives operating system services through the set of system calls. Earlier these calls were available in assembly language instructions but now a days these features are supported through high-level languages like C, Pascal etc., which replaces assembly language for system programming. The use of system calls in C or Pascal programs very much resemble pre-defined function or subroutine calls. As an example of how system calls are used, let us consider a simple program to copy data from one file to another. In an interactive system, the following system calls will be generated by the operating system:

1. Prompt messages for inputting two file names and reading it from terminal.
2. Open source and destination file.
3. Prompt error messages in case the source file cannot be open because it is protected against access or destination file cannot be created because there is already a file with this name.
4. Read the source file.
5. Write into the destination file.
6. Display status information regarding various Read/Write error conditions. For example, the program may find that the end of the file has been reached or that there was a hardware failure. The write operation may encounter various errors, depending upon the output device.
7. Close both files after the entire file is copied.

A user program takes heavy use of the operating system. All interaction between the program and its environment must occur as the result of requests from the program to the operating system.

Operating System Commands

Apart from system calls, users may interact with operating system directly by mews of operating system commands. For example, if you want to list files or sub-directories in MSDOS, you invoke dir command. In either case, the operating system acts as an interface between users and the hardware of a computer system. The fundamental goal of computer systems is to solve user problems. Towards this goal computer hardware is designed. Since the bare hardware alone is not very easy to use, programs (software) are developed. These programs require certain common operations, such as controlling peripheral devices. The command function of controlling and allocating resources are then brought together into one piece of software; the operating system. To see what operating systems are and what operating systems do, let us consider how they have evolved over the years. By tracing that evolution, we can identify the common elements of operating systems and examine how and why they have developed as they have.

OS and Hardware Development

OS development has gone hand-in-hand with hardware development:

— Interrupts drive data transfer (using multiple CPUs, one designed exclusively for I/O processing)

— Direct-memory-access (DMA) data transfer

— Hardware memory protection (to validate addresses)

— Hardware instruction protection (only "special" users can execute some machine instructions)

— Support for other interrupts: clock.

Old days: "busy wait", e.g. printing:

Pseudo-code:

1. check hardware switch to see if printer is ready for next character.
2. if switch not set, go to step 1.
3. send printer next char.

4. if more to print, go to step 1.
5. stop.

(steps 1 and 2 constitute the busy wait.) This same idea works for buffers as well as single characters: processor fills buffer (of maybe 512 characters), then uses busy wait for external device to signal that it has emptied the buffer. Though buffers are going to be more helpful if not using busy wait (why?) This waste of valuable CPU time (due to speed mismatches between CPU and external devices) is reduced by introduction of hardware interrupts.

Interrupt Handling

```
main memory
addr | inst
---+----
100 | br 200 Assume HW supports 16 interrupts
101 | br 300
102 | br 350
104 | br 550
. .
. .
. .
115 | br 950
. .
. .
. .
200 | xxxx <- SW to handle interrupt type 0
201 | xxxx
. .
. .
. .
300 | yyyy <- SW to handle interrupt type 1
. .
. .
. .
```

HW support

For each interrupt type (e.g. clock interrupt, IO completion, change to supervisor state, invalid instruction) the HW changes to program counter to a "prewired" memory address (in the above example, 100 for interrupt type 0, 101 for interrupt type 1, etc.) This results in the specific code to handle that interrupt being executed.

Operating Systems: Enhanced Features

As the need for better processing raised due to the increase in demand for better processing speed and efficiency the operating systems have been enhanced with extra features.

Serial Processing

Instructions and data are feeded into the computer by means of console switches or perhaps through a hexadecimal keyboard. Programs used to be started by loading the program counter register with the address of the first instruction of a program and its result used to be examined by the contents of various registers and memory locations of the machine. Therefore programming in this style caused a low utilisation of both users and machine.

With the advent of input output devices such as punched cards, paper tapes and language translators (compiler) assembler) bought significant computer system utilisation. Programs were coded into programming languages and then they are changed into object code (binary code) by translator and then automatically loaded into memory by a program called loader. Then the control is transferred to loaded program, the exhibition of a program begins and its result gets displayed or printed. Once in memory, the program may be re run with a different set of input data.

The process of development and preparation of a program in such environment is slow and cumbersome due to serial processing and numerous manual processing. In typical programming environment the following steps are performed.

— The source code is created in the editor by writing a user program.

— The source code is converted into binary code by the translator and

— The loader is called to load executable programs into main memory for execution. If syntax error is detected, the whole program must be restarted from the beginning.

— The next evolution was the replacement of card - decks with standard input output with some useful library programs, which were further linked with user program through system software called linker.

While there was a definite improvement overt machine language approach, the serial mode of operation is obviously not very efficient. This results in low utilisation of resources.

Batch Processing

During the time that tapes were being mounted or programmer was operating the console, the CPU sets idle. The next step in the logical evaluation of operating system was to automate the sequencing of operations involved in program execution and in the mechanical aspects of program development Programs With similar requirements were batched together and 11m through the computer as a group.

Example: Operator received one FORTRAN program, one COBOL program and another FORTRAN program. If he runs in that order ,he would have to load FORTRAN compiler tapes, then COBOL program Compiler and fmal1y FORTRAN compiler again. If he runs the two FORTRAN programs as a batch to save time. With batch processing utilisation of system resources has improved quite a bit What happens when a job is stopped?

When the job is stopped, the operator would have to notice that fact by observing the console, determines why the program is stopped and then loads the card reader or paper tape reader for the next job and restarts the computer.

Problems in. batch Processing

— The CPU sits idle when there is a job transition.

— Speed discrepancy between fast CPU and comparatively slow input/output devices such as card reader, printers.

The fist problem i.e. idle time of CPU can be overcomes by a small program called a resident monitor will be created, which resides always in the memory.

Resident Monitor: It acts according to the directives given by a programmer through control earth-which contain commands belonging to job control languages such as information like marking of job's beginning and ending, commands for loading and executing programs etc...

Example

—$COB - Execute the FORTAN compiler

—$JOB - First card of a job.

—$END - Last card of a job.

—$Load - Load program into memory.

—$RUN - Execute the user program.

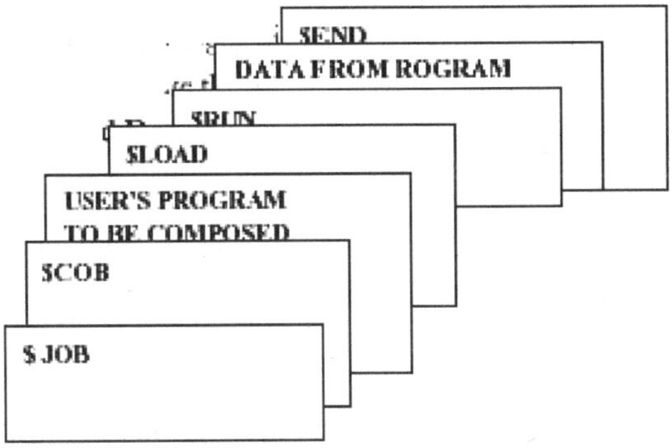

The second problem has been overcomes over the years through the technological improvement resulted in faster I/O devices. But CPU speed increased even faster. Therefore, the need was to increase the throughput and resource utilisation by overlapping I/O and processing operations. Dedicated I/O processors, peripheral controllers brought a major development. The development of Direct Memory Access(DMA) chip Was a major achievement, which directly transfer the entire block of data from its own memory buffer to main memory without intervention of CPU .DMA can transfer data between high speed I/O devices and main memory ,while the CPU is executing. CPU requires to be interrupted per block only by DMA. Apart from DMA, there are other two approaches to improve system performance by overlapping input/output and processing. These are:

1. Buffering
2. Spooling.

Buffering

It is a method of overlapping input/output and processing of a single job. The idea is quite simple. After data has been read and the CPU starts operating on it, the input device is instructed to begin the next input immediately. The CPU and the input device are both busy. The CPU can begin the processing of the newly read data, while the input device starts to read the following data. Similarly, this can be done for output In this case; the CPU creates data that is put into buffer until an output device can accept it.

What happens if the CPU is fast?

— In the case of input, the CPU finds an empty buffer and has to wait for the input device.

— In Case of output, the CPU can proceed at full speed until, eventually all system buffers are full. Then the CPU waits for the output device. This situation occurs with input/output bound jobs where the amount of input/output relation to computations very high. Since the CPU is faster than the input/output device, the speed of execution is controlled by the input/output device, not by the speed of the CPU.

Spooling

It stands for simultaneous peripheral operation on line. It is essentially use the disk as a large buffer for reading and for storing output files. Spooling allows CPU to overlap the input of one job with the computation and output of other jobs. Even in a simple system, the spooler may be reading the input of one job while printing the output of a different job. Compared to buffering approach spooling is better.

Multi-Programming

Buffering and spooling improve system performance by overlapping the input, output and computation of a single job, but both of them have limitations. A single user cannot always keep CPU or I/O devices busy at all times. Multiprogramming offers a more efficient approach to increase system performance. It refers to a computer system's ability to support more than one process (program) at the same time. Multiprocessing operating systems enable several programs to nm concurrently. This is a kind of parallel processing. More number of programs competing for system resources which lead to better utilisation of system resources. The idea is implemented as follows. The main memory of a system contains more than on program.

The operating system picks one of the programs and start executing. During execution process program 1 may need some I/O operation to complete. In a sequential Execution environment, the CPU would sit idle. In a Multiprogramming system, operating system will simply switch over the next program.

When that program needs to wait for some I/O operation, it switches over to program 3 and so OIL if there is no other new program left in the main memory, the CPU will pass its control back to the previous programs.

Compared to operating system which supports only sequential execution, multiprogramming system requires some form of CPU and memory management strategies. With each new generation of operating systems, we are introduced to new ways of thinking about how our computers work. To simplify things for the user, we must deploy a consistent interface in which they can do their work. It is equally important to extend this consistency to programmers, so they too can benefit. As an operating system ages, it gradually becomes burdened with a plethora of interfaces which break the simplicity of its original architecture. UNIX originally followed the "everything is a file" mantra, only to lose sight of that design with numerous task-specific APIs for transferring files FTP, HTTP, RCP, etc.), graphics (X11, svgalib), printers (lp, lpr), etc. Plan 9, introduced in 1989, and demonstrated how even a GUI can be represented as a set of files, revitalising the "everything is a file" idea.

Types of Operating System

Batch Operating System

As discussed earlier during batch processing environment it requires grouping of similar jobs which consist of programs, data and system commands. The suitability of this type of processing is in programs with large computation time with no need of user interaction/ involvement. Some examples of such programs include payroll, forecasting, statistical analysis and large scientific number crunching programs., Users are not required to wait while the job is being processed. They can submit their programs to operators and return later to collect them.

But it has two major disadvantages:

— Non-interactive environment and

— Off-line debugging.

Non-interactive environment: There are some difficulties with a batch system from the point of view of programmer or user. Batch operating systems allow little or no interaction between users and executing programs. The turn around time taken between job submission and job completion in batch operating system is very high. Users have no control over intermediate results of a program. This type, of arrangement does not create flexibility in software development.

The second disadvantage with this approach is that programs must be debugged which means a programmer cannot correct bugs the moment it occurs.

Process scheduling (i.e. allocation strategy for a process to a processor), memory management file management and I/O management in batch processing are quite simple. Jobs are typically processed in the order of submission, that is, in the first come, first served basis. Memory is usually divided into two areas. One of them is permanently fixed for containing operating system routines and the other part contains only user programs to be executed; when one Program is over, the new program is loaded into the same area. Since there is only one Program in the execution at a time, there is no competition for 110 devices, therefore, allocation and deallocation for 1/0 devices is very trivial. Access to files is also serial and there is hardly a need of Protection and file access control mechanism.

Multiprogramming Operating System

Multiprogramming operating systems compared to batch operating systems are fairly sophisticated. Multiprogramming has a significant potential for improving system throughput and resource utilisation with very minor differences. Different forms of multiprogramming operating system are multitasking, multiprocessor and multi-user operating systems. In this section, we will briefly discuss the main features and functions of these systems.

Multitasking operating systems

A running stale of a program is called a process or a task. A multitasking operating system (also called multiprocessing operating system) supports two or more active processes simultaneously. Multiprogramming operating system is operating system which, in addition to supporting multiple concurrent process (several processes in execution states simultaneously) allows the instruction and data from two or more separate processes to reside in primary memory simultaneously. Note that multiprogramming implies multiprocessing or multitasking operation, but multiprocessing operation (or multitasking) does not imply multiprogramming. Therefore, multitasking operation is one of the mechanism that multiprogramming operating system employs in managing the totality of computer related resources like CPU, memory and I/O devices.

The simplest form of multitasking is called serial multitasking or context switching. This is nothing more than stopping one temporarily to work on another. If you have used sidekick, then you have used serial multitasking. While a program is running, you decide that you want to use the calculator, so you pop it and use it. When you stop using the calculator, the Program continues running.

Multiuser operating system

Allow simultaneous access to a computer system through or more terminals. Although frequently associated with multiprogramming, multiuser operating system does, not imply multiprogramming or multitasking. A dedicated transaction processing system such as railway reservation system that hundreds of terminals under control of a single program is an example of multiuser operating system. On the other hand, general purpose time sharing systems incorporate features of both multiuser and multiprogramming operating system. Multiprocess operation without multiuser support can be found in the operating system of some advanced personnel computers and in real systems.

Time sharing system

It is a form of multiprogrammed Operating system which operates in an interactive mode with a quick response time. The user types a request to the computer through a keyboard. The computer processes it and a response (if any) is displayed on the user's terminal. A time sharing system allows the many users to simultaneously share the computer resources. Since each action or command in a time-shared system take a very small fraction of time, only a little CPU time is needed for each user. As the CPU switches rapidly from one user to another user, each user is given impression that he has his own computer, while it is actually one computer shared among many users.

Most time sharing system use time-slice (round robin) scheduling of CPU. In this approach, Programs are executed with rotating priority that increases during waiting and drops after the service is granted. In Order to prevent a program from monopolising the processor, a program executing longer than the system defined time-slice in interrupted by the operating system and placed at the end of the queue of waiting program. Memory management in time sharing system Provides for the protection and separation of user programs. Input/output management feature of timesharing system must be able to handle multiple users (terminals). However, the processing of terminals interrupts are not time critical due to the relative slow speed of terminals and users. As required by most multiuser environment allocation and deallocation of devices must be performed in a manner that preserves system integrity and provides for good performance.

The words multiprogramming, multiprocessing and multitasking are often confused. There are, of course, some distinctions between these similar, but distinct terms. The term multiprogramming refers to the situation in which a single CPU divides its time between more than one job. Time sharing is a special case of multiprogramming, where a single CPU serves a number of users at interactive terminals. In multiprocessing, multiple CPUs perform more than one job at one time. Multiprogramming and multiprocessing are not mutually exclusive. Some mainframes and super mini computers have multiple CPUs each of which can juggle several jobs. The term multitasking is described any system that runs or appears to run more than one application program one time. An effective multitasking environment must provide many services both to the user and to the application program it runs. The most important of these are resource management which divides the computers time, memory and peripheral devices among competing tasks and inter-process communication, which lets tasking coordinate their activities by exchanging information.

Real-time systems

It is another form of operating system which is used in environments where a large number of events mostly external to computer systems, must be accepted and processed in a short time or within certain deadlines. Examples of such applications are flight control, real time simulations etc. Real time systems are also frequently used in military application. A primary objective of real-time system is to provide quick response times. User convenience and resource utilisation are of secondary concern to real-time system. In the real-time system each process is assigned a certain level of priority according to the relative importance of the even it processes. The processor is normally allocated to the highest priority process among those which are ready to execute. Higher priority process usually preemptive execution of lower priority processes. This form of scheduling called, priority based pre-emptive scheduling, is used by a majority of real-time systems.

Memory Management: In real-time operating system there is a little swapping of program between primary and secondary memory. Most of die time, processes remain in primary memory in order to provide quick response, therefore, memory management in real-time system is less demanding compared to other types of multiprogramming system. On the other hand, processes in real- time system tend to cooperate closely thus providing feature for both protection and sharing of memory.

I/O Management: Time-critical device management is one of the main characteristics of real-time systems. It also provides sophisticated form of interrupt management and I/O buffering.

File Management: The primary objective of file management in real-time systems is usually the speed of access rather than efficient utilisation of secondary storage. In fact, some embedded real-time systems do not have secondary memory. However, where provided file management of real-time system must satisfy the same requirement as those found in time sharing and other multiprogramming systems.

Network Operating System

A network operating system is a collection of software and associated protocols that allow a set of autonomous computers which are interconnected by a computer network to be used together in a convenient and cost-effective manner. In a network operating system, the users are aware of existence of multiple computers and can log in to remote machines and copy files from one machine to another machine. Some of typical characteristics of network operating systems which make it different from distributed operating system are the followings:

— Each computer has its own private operating system instead of running part of a global system wide operating system.

— Each user normally works on his/her own system; using a different system requires some kind of remote login, instead of having the operating system dynamically allocate processes to CPUs.

— Users are typically aware of where each of their files are kept and must move file from one system to another with explicit file transfer commands instead of having file placement managed by the operating system.

The system has little or no fault tolerance; if 5% of the personnel computers crash, only 5% of the users is out of business.

Network operating system offers many capabilities including:

— Allowing users to access the various resources of the network hosts

— Controlling access so that only users in the proper authorisation are allowed to access particular resources.

— Making the use of remote resources appear to be identical to the use of local resources

??Providing up-to-the minute network documentation online.

The key issue that distinguishes a network operating system from a distributed one is how aware the users are of the fact that multiple machines are being used. This visibility occurs in three primary areas; file system, protection and program execution.

File system

The important issue in file system is related to how a file is placed (accessed) on one system from another in a network. Mere are two important approaches to this problem.

— Running a special file transfer program

— Specifying a path name

Running a special file transfer program: When connecting two or more systems together, the first issue that must be faced is how to access the rile system available on some other system. To deal with this issue user runs a special file transfer program that copies the needed remote file to the local machine, where they can then be accessed normally. Sometimes remote printing and mail is also handled this way. One of the best known examples of network that primarily support file transfer and mail via special programs is the UNIX's UUCP (user to user control program) program and its network USENET.

Path name specification: The second approach in this direction is that programs from one machine can open files on another machine by providing a path name telling where the file is located.

Execution location

Program execution is the third area in which machine boundaries are visible in network operating systems. When a user or a running program wants to create a new process, where is the process created? At least four schemes have been used thus far. The first of these is that the user simply says

"CREATE PROCESS" in one way or another, and specifies nothing about where. Depending on the implementation, this can be the best or worse Way to do it.

The second approach to process location is to allow users to run jobs on any machine by first logging in there. In this model, processes on different machines cannot communicate or exchange data, but a simple manual load balancing is possible. The third approach is a special command that the user types at a terminal to cause a program to be executed on a specific machine. A typical command might be "remote vax4 who" to run the who program on machine vax4. In M arrangement, the environment of the new process is the remote machine. In other words, if that process tries to read or write files from its current working directory, it will discover that its working directory is on the remote machine, and that files that were in the parent process's directory are no longer present. Similarly, files written in the working directory will appear on the remote machine, not the local one.

The fourth approach is to provide the "CREATE PROCESS" system call with a parameter specifying where to run the new process, possibly with a new system call for specifying the default site. As with the previous method, the environment will generally be the remote machine. In many cases, signals and other forms of inter-process communication between processes do not work properly among processes on different machines. Now let us see how file system protection and program execution are supported in distributed operating systems.

Distributed Operating System

A distributed operating system is one that looks to its users like an ordinary centralised operating system but runs on multiple independent CPUs. The key concept here is transparency. In other words, the use of multiple processors should be invisible to the user. Another way of expressing the same idea is to say that user views the system as virtual uniprocessor but not as a collection of distinct machines. In a true distributed system, users are not aware of where their programs are being run or where their files are residing; they should all be handled automatically and efficiently by the operating system.

Distributed operating systems have many aspects in common with centralised ones but they also differ in certain ways. Distributed operating system, for example, often allow programs to run on several processors at the same time, thus requiring more complex processor scheduling (scheduling refers to a set of policies and mechanisms built into the operating systems that controls the order in which the work to be done is completed) algorithms in order to achieve maximum utilisation of CPU's time.

Fault-tolerance is another area in which distributed operating systems are different. Distributed systems are considered to be more reliable than uniprocessor based system. They perform even if certain part of the hardware is malfunctioning. This additional feature, supported by distributed operating system has enormous implications for the operating system.

There are three important advantages in the design of distributed operating system:

1. *Major breakthrough in microprocessor technology* : Micro- processors have become very much powerful and cheap, compared with mainframes and minicomputers, so it has become attractive to think about designing large systems consisting of small processors. These distributed systems clearly have a price/performance advantages over more traditional systems.
2. *Incremental Growth :* The second advantage is that if there is a need of 10 per cent more computing power, one should just add 10 per cent more processors. System architecture is crucial to the type of system growth, however, since it is hard to give each user of a personal computer another 10 per cent.
3. *Reliability* : Reliability and availability can also be a big advantage; a few parts of the system can be down without disturbing people using the other parts; On the minus side, unless one is very careful, it is easy for the communication protocol overhead to become a major source of inefficiency.

Now let us see how file system, protection and program execution are supported in distributed operating system.

Distributed operating system supports a single global file system visible from all machines. When this method is used, there is one directory for executable programs (in UNIX, it is bin directory), one password file and so on. When a program wants to read the password file it does something like Open (*/etc/password", READ-ONLY) without reference to where the file is. It is upto the operating system to locate the file and arrange for transport of data as they are needed.

The convenience of having a single global name space is obvious. In addition, this approach means that operating system is free to move files around among machines to keep all the disks generally full and busy and that the system can maintain replicated copies of files if it chooses. When the user or program must specify the machine name, the system cannot decide on its own to move a file to a new machine because that would change the (user visible) name used to access the file. Thus in a network operating system, control over file placement must be done manually by the users, whereas in a distributed operating system it can be done automatically by the system itself.

In a true distributed system there is a unique UID for every user, and that UID should be valid on all machines without any mapping. In this way no protection problems arise on remote access to files; a remote access can be treated like a local access with the same UID. There is a difference between network operating system and distributed operating system in implementing protection issue. In networking operating system, there are various machines, each with its own user to UID mapping but in distributed operating system there is a single system wide mapping that is valid everywhere.

In the most distributed case the system chooses a CPU by looking at the processing load of the machine, location of file to be used etc. In the least distributed case, the system always run the process on one specific machine (usually the machine on which the user is logged in).

An important difference between network and distributed operating system is how they are implemented. A common way to realise a networking operating system is to put a layer of software on top of the native operating system of the individual machines. For example one could write a special library package that could intercept all the system calls and decide whether each one was local or remote. Although most system calls can be handled this way without modifying kernel.

Functions of an Operating System

When the power to a computer is turned on, the first program that runs is usually a set of instructions kept in the computer's read-only memory (ROM) that examines the system hardware to make sure everything is functioning properly. This power-on self test (POST) checks the CPU, memory, and basic inputoutput systems (BIOS) for errors and stores the result in a special memory location. Once the POST has successfully completed, the software loaded in ROM (sometimes called firmware) will begin to activate the computer's disk drives. In most modern computers, when the computer activates the hard disk drive, it finds the first piece of the operating system: the bootstrap loader.

The bootstrap loader is a small program that has a single function: It loads the operating system into memory and allows it to begin operation. In the most basic form, the bootstrap loader sets up the small driver programs that interface with and control the various hardware subsystems of the computer. It sets up the divisions of memory that hold the operating system, user information and applications. It establishes the data structures that will hold the myriad signals, flags and semaphores that are used to communicate within and between the subsystems and applications of the computer. Then it turns control of the computer over to the operating system. The operating system's tasks, in the most general sense, fall into six categories:

- Processor management
- Memory management
- Device management
- Storage management
- Application interface
- User interface

While there are some who argue that an operating system should do more than these six tasks, and some operating system vendors do build many more utility programs and auxiliary functions into their operating systems, these six tasks define the core of nearly all operating systems. Let's look at the tools the operating system uses to perform each of these functions.

Processor Management

The heart of managing the processor comes down to two related issues:

- Ensuring that each process and application receives enough of the processor's time to function properly.
- Using as many processor cycles for real work as is possible

The basic unit of software that the operating system deals with in scheduling the work done by the processor is either a process or a thread, depending on the operating system.

It's tempting to think of a process as an application, but that gives an incomplete picture of how processes relate to the operating system and hardware. The application you see (word processor or spreadsheet or game) is, indeed, a process, but that application may cause several other processes to begin, for tasks like communications with other devices or other computers. There are also numerous processes that run without giving you direct evidence that they ever exist. A process, then, is software that performs some action and can be controlled by a user, by other applications or by the operating system.

It is processes, rather than applications, that the operating system controls and schedules for execution by the CPU. In a single-tasking system, the schedule is straightforward. The operating system allows the application to begin running, suspending the execution only long enough to deal with interrupts and user input. Interrupts are special signals sent by hardware or software to the CPU. It's as if some part of the computer suddenly raised its hand to ask for the CPU's attention in a lively meeting. Sometimes the operating system will schedule the priority of processes so that interrupts are masked that is, the operating system will ignore the interrupts from some sources so that a particular job can be finished as quickly as possible. There are some interrupts (such as those from error conditions or problems with memory) that are so important that they can't be ignored. These non-maskable interrupts (NMIs) must be dealt with immediately, regardless of the other tasks at hand.

While interrupts add some complication to the execution of processes in a single-tasking system, the job of the operating system becomes much more complicated in a multi-tasking system. Now, the operating system must arrange the execution of applications so that you believe that there are several things happening at once. This is complicated because the CPU can only do one thing at a time. In order to give the appearance of lots of things happening at the same time, the operating system has to switch between different processes thousands of times a second. Here's how it happens.

— A process occupies a certain amount of RAM. It also makes use of registers, stacks and queues within the CPU and operating system memory space.

— When two processes are multi-tasking, the operating system allots a certain number of CPU execution cycles to one program.

— After that number of cycles, the operating system makes copies of all the registers, stacks and queues used by the processes, and notes the point at which the process paused in its execution.
— It then loads all the registers; stacks and queues used by the second process and allow it a certain number of CPU cycles.
— When those are complete, it makes copies of all the registers, stacks and queues used by the second program, and loads the first program.
— All of the information needed to keep track of a process when switching is kept in a data package called a process control block.

The process control block typically contains:

— An ID number that identifies the process
— Pointers to the locations in the program and its data where processing last occurred
— Register contents
— States of various flags and switches
— Pointers to the upper and lower bounds of the memory required for the process
— A list of files opened by the process
— The priority of the process
— The status of all I/O devices needed by the process

When the status of the process changes, from pending to active, for example, or from suspended to running, the information in the process control block must be used like the data in any other program to direct execution of the taskswitching portion of the operating system. This process swapping happens without direct user interference, and each process gets enough CPU cycles to accomplish its task in a reasonable amount of time. Trouble can come, though, if the user tries to have too many processes functioning at the same time. The operating system itself requires some CPU cycles to perform the saving and swapping of all the registers, queues and stacks of the application processes. If enough processes are started, and if the operating system hasn't been carefully designed, the system can begin to use the vast majority of its available CPU cycles to swap between processes rather than run processes. When this happens, it's called thrashing, and it usually requires some sort of direct user intervention to stop processes and bring order back to the system.

One way that operating-system designers reduce the chance of thrashing is by reducing the need for new processes to perform various tasks. Some operating systems allow for a "processlite," called a thread, that can deal with all the CPU-intensive work of a normal process, but generally does not deal with the various types of I/O and does not establish structures requiring the extensive process control block of a regular process. A process may start many threads or other processes, but a thread cannot start a process.

So far, all the scheduling we've discussed has concerned a single CPU. In a system with two or more CPUs, the operating system must divide the workload among the CPUs, trying to balance the demands of the required processes with the available cycles on the different CPUs. Asymmetric operating systems use one CPU for their own needs and divide application processes among the remaining CPUs. Symmetric operating systems divide themselves among the various CPUs, balancing demand versus CPU availability even when the operating system itself is all that's running. Even if the operating system is the only software with execution needs, the CPU is not the only resource to be scheduled. Memory management is the next crucial step in making sure that all processes run smoothly.

Memory and Storage Management

When an operating system manages the computer's memory, there are two broad tasks to be accomplished:

— Each process must have enough memory in which to execute, and it can neither run into the memory space of another process nor be run into by another process.

— The different types of memory in the system must be used properly so that each process can run most effectively.

The first task requires the operating system to set up memory boundaries for types of software and for individual applications. As an example, let's look at an imaginary system with 1 megabyte (1,000 kilobytes) of RAM. During the boot process, the operating system of our imaginary computer is designed to go to the top of available memory and then "back up" far enough to meet the needs of the operating system itself. Let's say that the operating system needs 300 kilobytes to run. Now, the operating system goes to the bottom of the pool of RAM and starts building up with the various driver software required to control the hardware subsystems of the computer. In our imaginary computer, the drivers take up 200 kilobytes. So after getting the operating system completely loaded, there are 500 kilobytes remaining for application processes. When applications begin to be loaded into memory, they are loaded in block sizes determined by the operating system. If the block size is 2 kilobytes, then every process that is loaded will be given a chunk of memory that is a multiple of 2 kilobytes in size. Applications will be loaded in these fixed block sizes, with the blocks starting and ending on boundaries established by words of 4 or 8 bytes.

These blocks and boundaries help to ensure that applications won't be loaded on top of one another's space by a poorly calculated bit or two. With that ensured, the larger question is what to do when the 500-kilobyte application space is filled.

In most computers, it's possible to add memory beyond the original capacity. For example, you might expand RAM from 1 to 2 megabytes. This works fine, but tends to be relatively expensive. It also ignores a fundamental fact of computing - most of the information that an application stores in memory is not being used at any given moment. A processor can only access memory one location at a time, so the vast majority of RAM is unused at any moment. Since disk space is cheap compared to RAM, then moving information in RAM to hard disk can greatly expand RAM space at no cost. This technique is called virtual memory management.

Disk storage is only one of the memory types that must be managed by the operating system, and is the slowest. Ranked in order of speed, the types of memory in a computer system are:

— *High-speed cache* - This is fast, a relatively small amount of memory that are available to the CPU through the fastest connections. Cache controllers predict which pieces of data the CPU will need next and pull it from main memory into high-speed cache to speed up system performance.

— *Main memory* - This is the RAM that you see measured in megabytes when you buy a computer.

— *Secondary memory* - This is most often some sort of rotating magnetic storage that keeps applications and data available to be used, and serves as virtual RAM under the control of the operating system.

The operating system must balance the needs of the various processes with the availability of the different types of memory, moving data in blocks (called pages) between available memory as the schedule of processes dictates.

Device Management

The path between the operating system and virtually all hardware not on the computer's motherboard goes through a special program called a driver. Much of a driver's function is to be the translator between the electrical signals of the hardware subsystems and the high-level programming languages of the operating system and application programs. Drivers take data that the operating system has defined as a file and translate them into streams of bits placed in specific locations on storage devices, or a series of laser pulses in a printer.

Because there are such wide differences in the hardware controlled through drivers, there are differences in the way that the driver programs function, but most are run when the device is required, and function much the same as any other process. The operating system will frequently assign high-priority blocks to drivers so that the hardware resource can be released and readied for further use as quickly as possible.

One reason that drivers are separate from the operating system is so that new functions can be added to the driver and thus to the hardware subsystems - without requiring the operating system itself to be modified, recompiled and redistributed. Through the development of new hardware device drivers, development often performed or paid for by the manufacturer of the subsystems rather than the publisher of the operating system, input/ output capabilities of the overall system can be greatly enhanced.

Managing input and output is largely a matter of managing queues and buffers, special storage facilities that take a stream of bits from a device, perhaps a keyboard or a serial port, hold those bits, and release them to the CPU at a rate slow enough for the CPU to cope with. This function is especially mportant when a number of processes are running and taking up processor time. The operating system will instruct a buffer to continue taking input from the device, but to stop sending data to the CPU while the process using the input is suspended.

Then, when the process needing input is made active once again, the operating system will command the buffer to send data. This process allows a keyboard or a modem to deal with external users or computers at a high speed even though there are times when the CPU can't use input from those sources. Managing all the resources of the computer system is a large part of the operating system's function and, in the case of realtime operating systems, may be virtually all the functionality required. For other operating systems, though, providing a relatively simple, consistent way for applications and humans to use the power of the hardware is a crucial part of their reason for existing.

Interfaces

Application Interface

Application program interfaces (APIs) let application programmers use functions of the computer and operating system without having to directly keep track of all the details in the CPU's operation. Let's look at the example of creating a hard disk file for holding data to see why this can be important. A programmer writing an application to record data from a scientific instrument might want to allow the scientist to specify the name of the file created. The operating system might provide an API function named MakeFile for creating files. When writing the program, the programmer would insert a line that looks like this: MakeFile [1, %Name, 2]. In this example, the instruction tells the operating system to create a file that will allow random access to its data (1), will have a name typed in by the user (%Name), and will be a size that varies depending on how much data is stored in the file (2). Now, let's look at what the operating system does to turn the instruction into action.

1. The operating system sends a query to the disk drive to get the location of the first available free storage location.
2. With that information, the operating system creates an entry in the file system showing the beginning and ending locations of the file, the name of the file, the file type, whether the file has been archived, which users have permission to look at or modify the file, and the date and time of the file's creation.
3. The operating system writes information at the beginning of the file that identifies the file, sets up the type of access possible and includes other information that ties the file to the application.

In all of this information, the queries to the disk drive and addresses of the beginning and ending point of the file are in formats heavily dependent on the manufacturer and model of the disk drive. Because the programmer has written her program to use the API for disk storage, she doesn't have to keep up with the instruction codes, data types, and response codes for every possible hard disk and tape drive.

The operating system, connected to drivers for the various hardware subsystems, deals with the changing details of the hardware - the programmer must simply write code for the API and trust the operating system to do the rest. APIs have become one of the most hotly contested areas of the computer industry in recent years. Companies realise that programmers using their API will ultimately translate into the ability to control and profit from a particular part of the industry. This is one of the reasons that so many companies have been willing to provide applications like readers or viewers to the public at no charge. They know consumers will request that programs take advantage of the free readers, and application companies will be ready to pay royalties to allow their software to provide the functions requested by the consumers.

User Interface

A User Interface (UI) brings structure to the interaction between a user and the computer. In the last decade, almost all development in user interfaces has been in the area of the graphical user interface (GUI), with two models, Apple's Macintosh and Microsoft's Windows, receiving most of the attention and gaining most of the market share. There are other user interfaces, some graphical and some not, for other operating systems.

Unix, for example, has user interfaces called shells that present a user interface more flexible and powerful than the standard operating system text-based interface. Programs such as the Korn Shell and the C Shell are text-based interfaces that add important utilities, but their main purpose is to make it easier for the user to manipulate the functions of the operating system.

There are also graphical user interfaces, such as X-Windows and Gnome, which make Unix and Linux more like Windows and Macintosh computers from the user's point of view. It's important to remember that in all of these examples, the user interface is a program or set of programs that sits as a layer above the operating system itself. The same thing is true, with somewhat different mechanisms, of both Windows and Macintosh operating systems. The core operating-system functions, the management of the computer system, lie in the kernel of the operating system. The display manager is separate, though it may be tied tightly to the kernel beneath. The ties between the operating-system kernel and the user interface, utilities and other software define many of the differences in operating systems today, and will further define them in the future.

Operating Systems Structure

Since you all know operating system is a very large and complex software, it must be engineered carefully, if it is to be functioned properly and to be modified easily. It should be developed as a collection of several smaller modules with carefully defined inputs, outputs and functions rather than a single piece of software. Let us now examine different types of operating systems.

Simple Structure

This type of structures is not well defined .Such type of operating system are as :

— Small

— Simple.

— Limited systems.

— Grew beyond their scope.

Example 1

MS-DOS was written to provide the most functionality in the least space.

— not divided into modules

— Although MS-DOS has some structure, its interfaces and levels of functionality are not well separated.

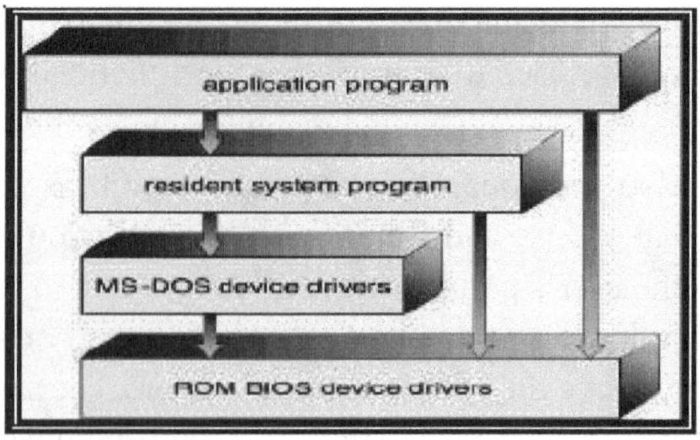

Figure 1. MS-DOS Layer Structure

Layered Approach

The operating system architecture based on layered approach is divided into a number of layers (levels), each built on top of lower layers. The bottom layer (layer 0), is the hardware; the highest (layer N) is the user interface. Higher level layers uses the functions (operations) and services of only lower-level layers.

Advantages

— As the system is divided into layers/modules verification and debugging of the system is easy/simple.

— Easy to find and rectify the errors ,as we can find in which layer the error has occurred.

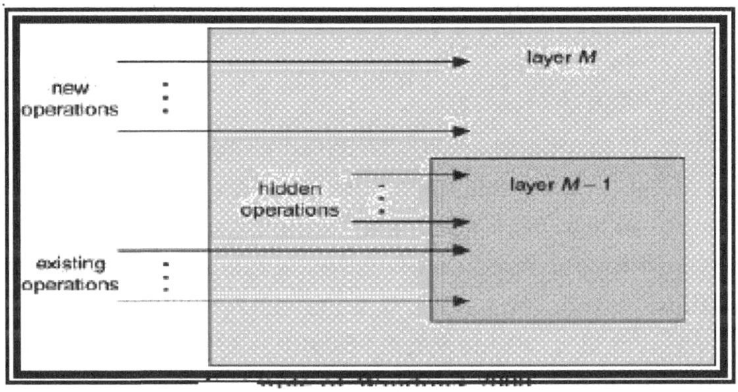

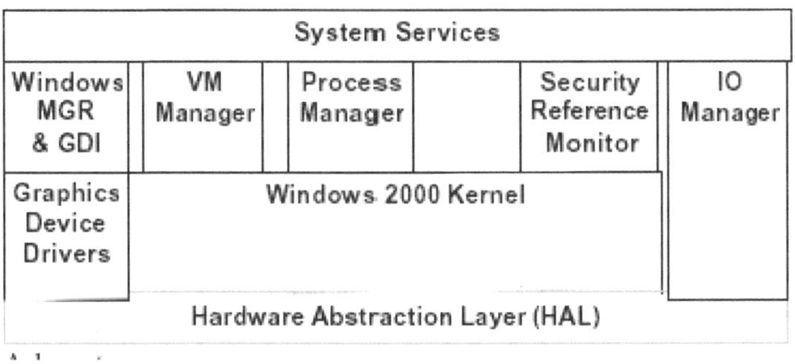

Figure 2. Example of Windows 2000.

— A layer need now the operations implemented by the lower layer, thus hides the existence of certain data structures, operations and hardware.

Difficulties for designing layered approach

— The layered approach involves careful definition of the layers, since the higher level layer uses the services of the lower level layer.

— For Example, the device driver for the secondary memory must be a lower level than memory management routines since memory management requires the ability to use the backing store.

— This approach tends to be less efficient than others. For example, When a user program executes an I/O operation, it executes a system call that is traped to the I/O Layer, which in turn calls the memory management layer, which in turn calls the CPU-scheduling layer, which is then passed to the hardware. At each layer , the parameters may be modified, data may need to be passed, and so on. Each layer adds overhead to the system call, thus system call takes longer time when compared to a non-layered system.

OS/2 Operating System Layer

OS/2 is the descendant of MS-DOS adds additional features such as multitasking and dual-mode operation and other addition features. In OS/2 fewer layers with more functionality are designed, providing most of the advantages of modularised code while avoiding the difficult problems of layer definition and interaction. The advantage in this type of operating system is ,direct access to low-level facilities is not allowed, providing the operating system with more control over the hardware and more knowledge of which resources each user program is using.

Kernel Approach

The kernel is the heart of the operating system. It is the part of operating system which directly makes interface with hardware system. When the system is booted, the kernel is read into the memory.

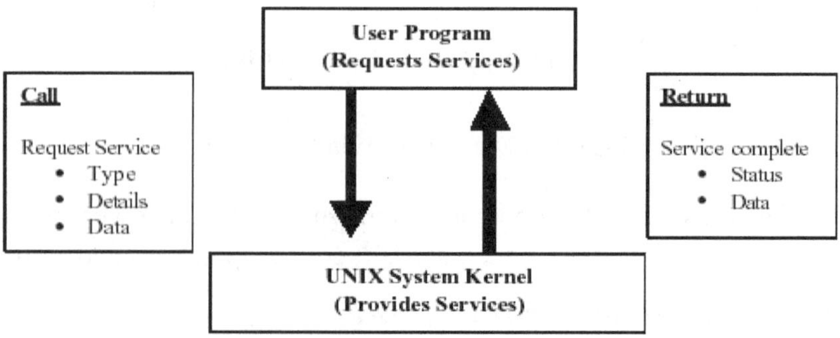

Figure 3. Making a system call

Kernel stays in memory while the system is running. Its main functions are:

— To provide a mechanism for creation and deletion of processes.

— To provide process scheduling, memory management and I/ O management.

— To provide mechanism for synchronisation of process so that processes synchronise their actions.

— To provide mechanism for interprocess communication.

The UNIX operating system is based on kernel approach. It consists of two parts: (a) Kernel and (b) System Program.

Programs and commands call on the kernel's services. The kernel, in turn, consults its data tables as it schedules user's programs, allocates resources to the program, and manages the low-level exchange of data with the computer's hardware. For example, when a program requests file services, the program gives the kernel a system call.

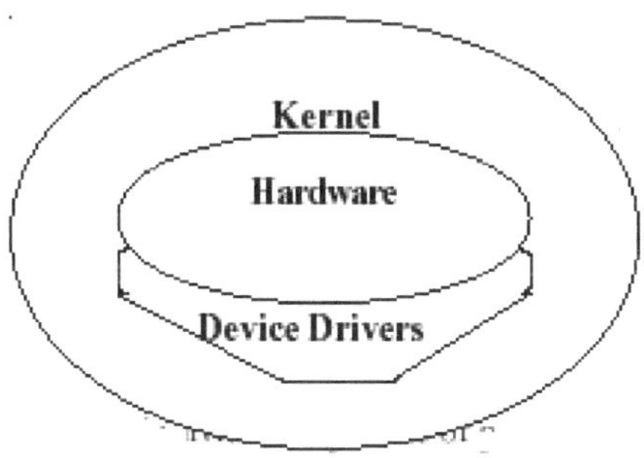

Figure 4. Kernel Approach

The kernel oversees the accessing of the disk drive where the file resides. The kernel gets the data and transfers it to the buffer. The data is then picked up by the parts of the kernel can be configured to accommodate variations in hardware .The kernel contains a changeable set of device drivers to accommodate numerous devices. With in the kernel, individual segments of programs or routines, carry out the kernel's work. The routines allocate memory resources, schedule CPU time, and manage access to system resources. The kernel also monitors the system for error conditions and hardware problems. At higher level, the routines provide programs with entry points to kernel services. All UNIX programs use the kernel's system call.

Virtual Machine

A virtual machine takes the layered approach to its logical conclusion. It treats hardware and the operating system kernel as though they were all hardware. A virtual machine provides an interface identical to the underlying bare hardware. The operating system creates the illusion of multiple processes, each executing on its own processor with its own (virtual) memory.

— The resources of the physical computer are shared to create the virtual machines.

— CPU scheduling can create the appearance that users have their own processor.

— Spooling and a file system can provide virtual card readers and virtual line printers.

— A normal user time-sharing terminal serves as the virtual machine operator's console.

— Protection is excellent, but no sharing possible.

— Virtual privileged instructions are trapped.

— Useful for running different OS simultaneously on the same machine.

Virtual machine multiprogramming

In conventional multiprogramming processes are allocated a portion of the real machines resources. The same machine resources are distributed among several processes.

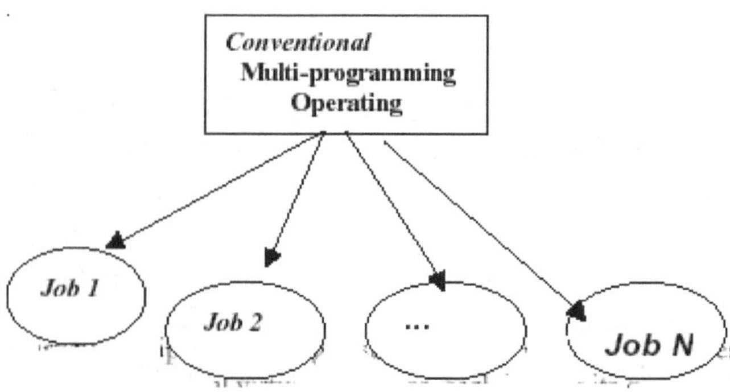

Figure 5. Conventional multiprogramming

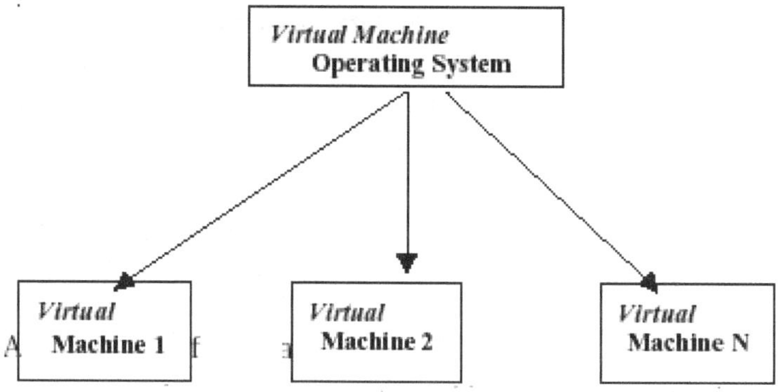

Figure 6. Virtual machine multiprogramming

In virtual multiprogramming system, a single real machine gives an illusion of several virtual machines, each having its own virtual processor, storage and I/O devices possibly with much larger capacity. Process scheduling can be used to share the CPU and make it appear that user's have their own processors. Virtual memory organisation technique can create illusion of very large memory for program execution.

Advantages of virtual machines

— The virtual-machine concept provides complete protection of system resources since each virtual machine is isolated from all other virtual machines.

— A virtual-machine system is a perfect vehicle for operatingsystems research and development. System development is done on the virtual machine, instead of on a physical machine and so does not disrupt normal system operation.

— The higher degree of separation between independent virtual machine aids in ensuring aids privacy and security.

Disadvantages of virtual machines

— Permits no direct sharing of resources.

— The virtual machine concept is difficult to implement due to the effort required to provide an exact duplicate to the underlying machine.

Client-Server Model

A trend in the modern operating system is to take moving code up into the higher layers even further, and remove as much as possible from the operating system, leaving a minimal kernel. The usual approach is to implement most of the operating system functions in user processes. To request a service, such as reading a block of a file, a user process sends the request to a server process, which then does the work and sends back the answer. In this model, the kernel does all the communication between the clients and servers by splitting the operating system into parts, each of which only handles:

— File service

— Process service

— Terminal service

— Memory service.

This way, each part becomes small and manageable. Furthermore, because all the servers run as user-mode processes, and not in kernel mode, they do not have direct access to the hardware. As a consequence, if a bug in the file server is triggered, the file service may crash, but this will not usually bring the whole system down.

Advantages

It is adaptable in distributed systems. If a client communicates with a server by sending it messages, the client need not know whether the message is handled locally in its own machine, or whether it was sent across a network to a server on a remote machine. As far as the client is concerned, the same thing happens in both cases, request was sent and a reply came back.

Message from Client to Server

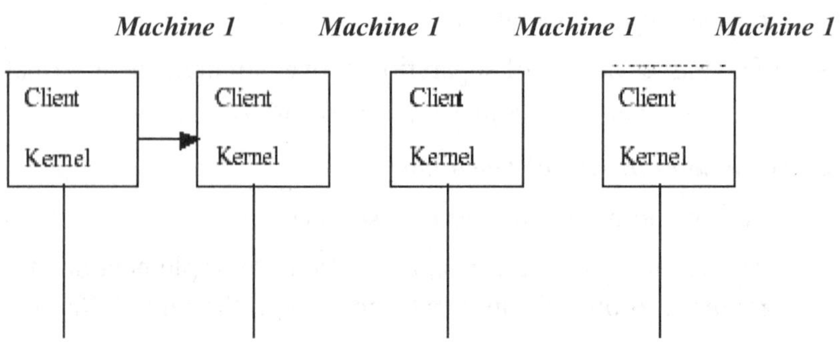

Figure 7. Message from Client to Server

The figure above of a kernel that handles only the transport of messages from clients to servers and back is not completely realistic. Some operating system functions (such as loading commands into the physical I/O device registers) are difficult, if not impossible, to do from user-space programs. Here are two ways of dealing with this problem.

— Is to have some critical server processes (eg:- I/O device drivers) actually run in kernel mode, with complete access to all the hardware, but still communicate with other processes using the normal message mechanism.

— Is to build a minimal amount of mechanism into the kernel, but leave the policy decisions up to servers in user space. For example, the kernel might recognise that a message sent to a certain special address means to take the contents of that message and load in into the I/O device registers for some disk, to start a disk read. In this example, the kernel would not even inspect the bytes in the message to see if they were valid or meaningful; it would just blindly copy them into the disk's device registers. The split between mechanism and policy is an important concept; it occurs again and again in operating Systems in various contexts.

EXAMPLES OF OPERATING SYSTEMS

UNIX and UNIX-like Operating Systems

Unix was originally written in assembly language. Ken Thompson wrote B, mainly based on BCPL, based on his experience in the MULTICS project. B was replaced by C, and Unix, rewriten in C, developed into a large, complex family of inter-related operating systems which have been influential in every modern operating system.

The UNIX-like family is a diverse group of operating systems, with several major sub-categories including System V, BSD, and Linux. The name "UNIX" is a trademark of The Open Group which licenses it for use with any operating system that has been shown to conform to their definitions. "UNIX-like" is commonly used to refer to the large set of operating systems which resemble the original UNIX.

Unix-like systems run on a wide variety of computer architectures. They are used heavily for servers in business, as well as workstations in academic and engineering environments. Free UNIX variants, such as Linux and BSD, are popular in these areas.

Four operating systems are certified by the The Open Group (holder of the Unix trademark) as Unix. HP's HP-UX and IBM's AIX are both descendants of the original System V Unix and are designed to run only on their respective vendor's hardware. In contrast, Sun Microsystems's Solaris Operating System can run on multiple types of hardware, including x86 and Sparc servers, and PCs. Apple's Mac OS X, a replacement for Apple's earlier (non-Unix) Mac OS, is a hybrid kernel-based BSD variant derived from NeXTSTEP, Mach, and FreeBSD.

Unix interoperability was sought by establishing the POSIX standard. The POSIX standard can be applied to any operating system, although it was originally created for various Unix variants.

BSD and its Descendants

A subgroup of the Unix family is the Berkeley Software Distribution family, which includes FreeBSD, NetBSD, and OpenBSD. These operating systems are most commonly found on webservers, although they can also function as a personal computer OS. The Internet owes much of its existence to BSD, as many of the protocols now commonly used by computers to connect, send and receive data over a network were widely implemented and refined in BSD. The world wide web was also first demonstrated on a number of computers running an OS based on BSD called NextStep.

BSD has its roots in Unix. In 1974, University of California, Berkeley installed its first Unix system. Over time, students and staff in the computer science department there began adding new programs to make things easier, such as text editors. When Berkely received new VAX computers in 1978 with Unix installed, the school's undergraduates modified Unix even more in order to take advantage of the computer's hardware possibilities. The Defense Advanced Research Projects Agency of the US Department of Defense took interest, and decided to fund the project. Many schools, corporations, and government organizations took notice and started to use Berkeley's version of Unix instead of the official one distributed by AT&T.

Steve Jobs, upon leaving Apple Inc. in 1985, formed NeXT Inc., a company that manufactured high-end computers running on a variation of BSD called NeXTSTEP. One of these computers was used by Tim Berners-Lee as the first webserver to create the World Wide Web.

Developers like Keith Bostic encouraged the project to replace any non-free code that originated with Bell Labs. Once this was done, however, AT&T sued. Eventually, after two years of legal disputes, the BSD project came out ahead and spawned a number of free derivatives, such as FreeBSD and NetBSD.

OS X

Mac OS X is a line of open core graphical operating systems developed, marketed, and sold by Apple Inc., the latest of which is pre-loaded on all currently shipping Macintosh computers. Mac OS X is the successor to the original Mac OS, which had been Apple's primary operating system since 1984. Unlike its predecessor, Mac OS X is a UNIX operating system built on technology that had been developed at NeXT through the second half of the 1980s and up until Apple purchased the company in early 1997. The operating system was first released in 1999 as Mac OS X Server 1.0, with a desktop-oriented version (Mac OS X v10.0 "Cheetah") following in March 2001. Since then, six more distinct "client" and "server" editions of Mac OS X have been released, the most recent being OS X 10.8 "Mountain Lion", which was first made available on February 16, 2012 for developers, and was then released to the public on July 25, 2012. Releases of Mac OS X are named after big cats.

The server edition, Mac OS X Server, is architecturally identical to its desktop counterpart but usually runs on Apple's line of Macintosh server hardware. Mac OS X Server includes work group management and administration software tools that provide simplified access to key network services, including a mail transfer agent, a Samba server, an LDAP server, a domain name server, and others. In Mac OS X v10.7 Lion, all server aspects of Mac OS X Server have been integrated into the client version.

Linux and GNU

Linux (or GNU/Linux) is a Unix-like operating system that was developed without any actual Unix code, unlike BSD and its variants. Linux can be used on a wide range of devices from supercomputers to wristwatches. The Linux kernel is released under an open source license, so anyone can read and modify its code. It has been modified to run on a large variety of electronics. Although estimates suggest that Linux is used on 1.82% of all personal computers, it has been widely adopted for use in servers and embedded systems (such as cell phones). Linux has superseded Unix in most places, and is used on the 10 most powerful supercomputers in the world. The Linux kernel is used in some popular distributions, such as Red Hat, Debian, Ubuntu, Linux Mint and Google's Android.

The GNU project is a mass collaboration of programmers who seek to create a completely free and open operating system that was similar to Unix but with completely original code. It was started in 1983 by Richard Stallman, and is responsible for many of the parts of most Linux variants. Thousands of pieces of software for virtually every operating system are licensed under the GNU General Public License. Meanwhile, the Linux kernel began as a side project of Linus Torvalds, a university student from Finland. In 1991, Torvalds began work on it, and posted information about his project on a newsgroup for computer students and programmers. He received a wave of support and volunteers who ended up creating a full-fledged kernel. Programmers from GNU took notice, and members of both projects worked to integrate the finished GNU parts with the Linux kernel in order to create a full-fledged operating system.

Google Chrome OS

Chrome is an operating system based on the Linux kernel and designed by Google. Since Chrome OS targets computer users who spend most of their time on the Internet, it is mainly a web browser with no ability to run applications. It relies on Internet applications (or Web apps) used in the web browser to accomplish tasks such as word processing and media viewing, as well as online storage for storing most files.

Microsoft Windows

Microsoft Windows is a family of proprietary operating systems designed by Microsoft Corporation and primarily targeted to Intel architecture based computers, with an estimated 88.9 percent total usage share on Web connected computers. The newest version is Windows 8 for workstations and Windows Server 2012 for servers. Windows 7 recently overtook Windows XP as most used OS.

Microsoft Windows originated in 1985 as an operating environment running on top of MS-DOS, which was the standard operating system shipped on most Intel architecture personal computers at the time. In 1995, Windows 95 was released which only used MS-DOS as a bootstrap. For backwards compatibility, Win9x could run real-mode MS-DOS and 16 bits Windows 3.x drivers. Windows Me, released in 2000, was the last version in the Win9x family. Later versions have all been based on the Windows NT kernel. Current versions of Windows run on IA-32 and x86-64 microprocessors, although Windows 8 will support ARM architecture. In the past, Windows NT supported non-Intel architectures.

Server editions of Windows are widely used. In recent years, Microsoft has expended significant capital in an effort to promote the use of Windows as a server operating system. However, Windows' usage on servers is not as widespread as on personal computers, as Windows competes against Linux and BSD for server market share.

Others

There have been many operating systems that were significant in their day but are no longer so, such as AmigaOS; OS/2 from IBM and Microsoft; Mac OS, the non-Unix precursor to Apple's Mac OS X; BeOS; XTS-300; RISC OS; MorphOS and FreeMint. Some are still used in niche markets and continue to be developed as minority platforms for enthusiast communities and specialist applications. OpenVMS formerly from DEC, is still under active development by Hewlett-Packard. Yet other operating systems are used almost exclusively in academia, for operating systems education or to do research on operating system concepts. A typical example of a system that fulfills both roles is MINIX, while for example Singularity is used purely for research.

References

Auslander, Marc A.; Larkin, David C.; Scherr, Allan L. 1981. *The evolution of the MVS Operating System*. IBM J. Research & Development.

Bic, Lubomur F.; Shaw, Alan C. 2003. *Operating Systems*. Pearson: Prentice Hall.

Deitel, Harvey M.; Deitel, Paul; Choffnes, David. *Operating Systems*. Pearson/Prentice Hall.

Silberschatz, Avi; Galvin, Peter; Gagne, Greg 2008. *Operating Systems Concepts*. John Wiley & Sons.

Stallings, William 2008. *Computer Organization & Architecture*. New Delhi: Prentice-Hall of India Private Limited. p. 267.

4

Distributed Systems

Distributed computing is a field of computer science that studies distributed systems. A distributed system consists of multiple autonomous computers that communicate through a computer network. The computers interact with each other in order to achieve a common goal. A computer program that runs in a distributed system is called a distributed program, and distributed programming is the process of writing such programs.

Distributed computing also refers to the use of distributed systems to solve computational problems. In distributed computing, a problem is divided into many tasks, each of which is solved by one or more computers, which communicate with each other by message passing.

The word distributed in terms such as "distributed system", "distributed programming", and "distributed algorithm" originally referred to computer networks where individual computers were physically distributed within some geographical area. The terms are nowadays used in a much wider sense, even referring to autonomous processes that run on the same physical computer and interact with each other by message passing. While there is no single definition of a distributed system, the following defining properties are commonly used:

- There are several autonomous computational entities, each of which has its own local memory.
- The entities communicate with each other by message passing.

In this chapter, the computational entities are called computers or nodes.

A distributed system may have a common goal, such as solving a large computational problem. Alternatively, each computer may have its own user with individual needs, and the purpose of the distributed system is to coordinate the use of shared resources or provide communication services to the users.

Other typical properties of distributed systems include the following:

- The system has to tolerate failures in individual computers.
- The structure of the system (network topology, network latency, number of computers) is not known in advance, the system may consist of different kinds of computers and network links, and the system may change during the execution of a distributed program.
- Each computer has only a limited, incomplete view of the system. Each computer may know only one part of the input.

Parallel and Distributed Computing

Distributed systems are groups of networked computers, which have the same goal for their work. The terms "concurrent computing", "parallel computing", and "distributed computing" have a lot of overlap, and no clear distinction exists between them. The same system may be characterised both as "parallel" and "distributed"; the processors in a typical distributed system run concurrently in parallel. Parallel computing may be seen as a particular tightly coupled form of distributed computing, and distributed computing may be seen as a loosely coupled form of parallel computing. Nevertheless, it is possible to roughly classify concurrent systems as "parallel" or "distributed" using the following criteria:

- In parallel computing, all processors may have access to a shared memory to exchange information between processors.
- In distributed computing, each processor has its own private memory (distributed memory). Information is exchanged by passing messages between the processors.

The figure on the right illustrates the difference between distributed and parallel systems. Figure (1a) is a schematic view of a typical distributed system; as usual, the system is represented as a network topology in which each node is a computer and each line connecting the nodes is a communication link. Figure (1b) shows the same distributed system in more detail: each computer has its own local memory, and information can be exchanged only by passing messages from one node to another by using the available communication links. Figure (1c) shows a parallel system in which each processor has a direct access to a shared memory.

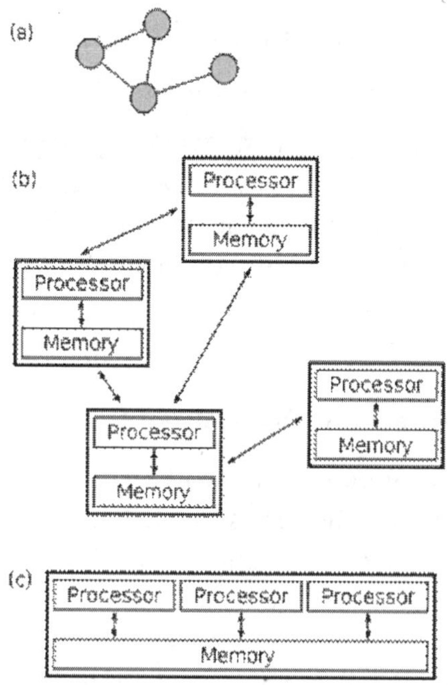

Figure 1. Parallel and distributed computing

The situation is further complicated by the traditional uses of the terms parallel and distributed algorithm that do not quite match the above definitions of parallel and distributed systems; see the section Theoretical foundations below for more detailed discussion. Nevertheless, as a rule of thumb, high-performance parallel computation in a shared-memory multiprocessor uses parallel algorithms while the coordination of a large-scale distributed system uses distributed algorithms.

Development

The use of concurrent processes that communicate by message-passing has its roots in operating system architectures studied in the 1960s. The first widespread distributed systems were local-area networks such as Ethernet that was invented in the 1970s.

ARPANET, the predecessor of the Internet, was introduced in the late 1960s, and ARPANET e-mail was invented in the early 1970s. E-mail became the most successful application of ARPANET, and it is probably the earliest example of a large-scale distributed application. In addition to ARPANET, and its successor, the Internet, other early worldwide computer networks included Usenet and FidoNet from 1980s, both of which were used to support distributed discussion systems.

The study of distributed computing became its own branch of computer science in the late 1970s and early 1980s. The first conference in the field, Symposium on Principles of Distributed Computing (PODC), dates back to 1982, and its European counterpart International Symposium on Distributed Computing (DISC) was first held in 1985.

Principles of Distributed Systems

There are several key principals underlying all distributed systems. As such, any distributed system can be described based on how those principles apply to that system.

— Concurrency
— Communication
— Synchronisation
— Replication and Consistency
— Naming
— Fault Tolerance
— Security

Paradigms

Most middleware systems (and, therefore, most distributed systems) are usually based on a particular paradigm, or model, for describing distribution and communication. Some of these paradigms are:

— Shared memory
— Distributed Computing
— Distributed objects
— Distributed file -system
— Shared documents

— Distributed coordination
— Agents

THEORETICAL FOUNDATIONS

Models

Many tasks that we would like to automate by using a computer are of question–answer type: we would like to ask a question and the computer should produce an answer. In theoretical computer science, such tasks are called computational problems. Formally, a computational problem consists of instances together with a solution for each instance. Instances are questions that we can ask, and solutions are desired answers to these questions.

Theoretical computer science seeks to understand which computational problems can be solved by using a computer (computability theory) and how efficiently (computational complexity theory). Traditionally, it is said that a problem can be solved by using a computer if we can design an algorithm that produces a correct solution for any given instance. Such an algorithm can be implemented as a computer program that runs on a general-purpose computer: the program reads a problem instance from input, performs some computation, and produces the solution as output. Formalisms such as random access machines or universal Turing machines can be used as abstract models of a sequential general-purpose computer executing such an algorithm.The field of concurrent and distributed computing studies similar questions in the case of either multiple computers, or a computer that executes a network of interacting processes: which computational problems can be solved in such a network and how efficiently? However, it is not at all obvious what is meant by "solving a problem" in the case of a concurrent or distributed system: for example, what is the task of the algorithm designer, and what is the concurrent or distributed equivalent of a sequential general-purpose computer?

The discussion below focuses on the case of multiple computers, although many of the issues are the same for concurrent processes running on a single computer.

Three viewpoints are commonly used:

Parallel algorithms in shared-memory model

- All computers have access to a shared memory. The algorithm designer chooses the program executed by each computer.
- One theoretical model is the parallel random access machines (PRAM) that are used. However, the classical PRAM model assumes synchronous access to the shared memory.
- A model that is closer to the behavior of real-world multiprocessor machines and takes into account the use of machine instructions, such as Compare-and-swap (CAS), is that of asynchronous shared memory. There is a wide body of work on this model, a summary of which can be found in the literature.

Parallel algorithms in message-passing model

- The algorithm designer chooses the structure of the network, as well as the program executed by each computer.
- Models such as Boolean circuits and sorting networks are used. A Boolean circuit can be seen as a computer network: each gate is a computer that runs an extremely simple computer program. Similarly, a sorting network can be seen as a computer network: each comparator is a computer.

Distributed algorithms in message-passing model

- The algorithm designer only chooses the computer program. All computers run the same program. The system must work correctly regardless of the structure of the network.
- A commonly used model is a graph with one finite-state machine per node.

In the case of distributed algorithms, computational problems are typically related to graphs. Often the graph that describes the structure of the computer network is the problem instance. This is illustrated in the following example.

Complexity Measures

In parallel algorithms, yet another resource in addition to time and space is the number of computers. Indeed, often there is a trade-off between the running time and the number of computers: the problem can be solved faster if there are more computers running in parallel. If a decision problem can be solved in polylogarithmic time by using a polynomial number of processors, then the problem is said to be in the class NC. The class NC can be defined equally well by using the PRAM formalism or Boolean circuits – PRAM machines can simulate Boolean circuits efficiently and vice versa.

In the analysis of distributed algorithms, more attention is usually paid on communication operations than computational steps. Perhaps the simplest model of distributed computing is a synchronous system where all nodes operate in a lockstep fashion. During each communication round, all nodes in parallel (1) receive the latest messages from their neighbours, (2) perform arbitrary local computation, and (3) send new messages to their neighbours. In such systems, a central complexity measure is the number of synchronous communication rounds required to complete the task.

This complexity measure is closely related to the diameter of the network. Let D be the diameter of the network. On the one hand, any computable problem can be solved trivially in a synchronous distributed system in approximately 2D communication rounds: simply gather all information in one location (D rounds), solve the problem, and inform each node about the solution (D rounds).

On the other hand, if the running time of the algorithm is much smaller than D communication rounds, then the nodes in the network must produce their output without having the possibility to obtain information about distant parts of the network. In other words, the nodes must make globally consistent decisions based on information that is available in their local neighbourhood. Many distributed algorithms are known with the running time much smaller than D rounds, and understanding which problems can be solved by such algorithms is one of the central research questions of the field.

Other Problems

Traditional computational problems take the perspective that we ask a question, a computer (or a distributed system) processes the question for a while, and then produces an answer and stops. However, there are also problems where we do not want the system to ever stop. Examples of such problems include the dining philosophers problem and other similar mutual exclusion problems. In these problems, the distributed system is supposed to continuously coordinate the use of shared resources so that no conflicts or deadlocks occur.

There are also fundamental challenges that are unique to distributed computing. The first example is challenges that are related to fault-tolerance. Examples of related problems include consensus problems, Byzantine fault tolerance, and self-stabilisation.

A lot of research is also focused on understanding the asynchronous nature of distributed systems:

- Synchronizers can be used to run synchronous algorithms in asynchronous systems.
- Logical clocks provide a causal happened-before ordering of events.
- Clock synchronization algorithms provide globally consistent physical time stamps.

Architecture

Various hardware and software architectures are used for distributed computing. At a lower level, it is necessary to interconnect multiple CPUs with some sort of network, regardless of whether that network is printed onto a circuit board or made up of loosely coupled devices and cables. At a higher level, it is necessary to interconnect processes running on those CPUs with some sort of communication system.

Distributed programming typically falls into one of several basic architectures or categories: client–server, 3-tier architecture, n-tier architecture, distributed objects, loose coupling, or tight coupling.

- Client–server: Smart client code contacts the server for data then formats and displays it to the user. Input at the client is committed back to the server when it represents a permanent change.
- 3-tier architecture: Three tier systems move the client intelligence to a middle tier so that stateless clients can be used. This simplifies application deployment. Most web applications are 3-Tier.
- n-tier architecture: n-tier refers typically to web applications which further forward their requests to other enterprise services. This type of application is the one most responsible for the success of application servers.
- highly coupled (clustered): refers typically to a cluster of machines that closely work together, running a shared process in parallel. The task is subdivided in parts that are made individually by each one and then put back together to make the final result.

- Peer-to-peer: an architecture where there is no special machine or machines that provide a service or manage the network resources. Instead all responsibilities are uniformly divided among all machines, known as peers. Peers can serve both as clients and servers.
- Space based: refers to an infrastructure that creates the illusion (virtualization) of one single address-space. Data are transparently replicated according to application needs. Decoupling in time, space and reference is achieved.

Another basic aspect of distributed computing architecture is the method of communicating and coordinating work among concurrent processes. Through various message passing protocols, processes may communicate directly with one another, typically in a master/slave relationship. Alternatively, a "database-centric" architecture can enable distributed computing to be done without any form of direct inter-process communication, by utilizing a shared database.

Properties of distributed systems

So far the focus has been on designing a distributed system that solves a given problem. A complementary research problem is studying the properties of a given distributed system.

The halting problem is an analogous example from the field of centralised computation: we are given a computer program and the task is to decide whether it halts or runs forever. The halting problem is undecidable in the general case, and naturally understanding the behaviour of a computer network is at least as hard as understanding the behaviour of one computer.

However, there are many interesting special cases that are decidable. In particular, it is possible to reason about the behaviour of a network of finite-state machines. One example is telling whether a given network of interacting (asynchronous and non-deterministic) finite-state machines can reach a deadlock. This problem is PSPACE-complete, i.e., it is decidable, but it is not likely that there is an efficient (centralised, parallel or distributed) algorithm that solves the problem in the case of large networks.

Communication in Distributed Systems

Inter process communication is at the heart of all distributed systems. Communication in distributed systems is always based on low-level message passing as offered by the underlying network.

Layered Protocols

Due to the absence of shared memory, all communication in distributed systems in based on exchanging(low level) messages. When process A wants to communicate with process B, it first builds a message in its own address space. Then it executes a system call that causes the operating system to send the message over the network to B. To make it easier to deal with the numerous levels and issues involved in communication, the International Standard Organisation (ISO) developed a reference model that clearly identifies the various levels involved, gives them standard names, and points out which level should do which job. This model is called OSI model.

Higher -Level Protocols

Higher level protocols are at these layers:

— Application layer

— Presentation layer

— Session layer

Lower -Level Protocols

Lower layer protocols are at these layers:

— Physical Layer

— Data Link Layer

— Network Layer

— Transport Layer

Client-Server Interaction

Client-Server interaction in distributed systems is often done using the transport protocols of the underlying network. With the increasing popularity of the Internet, it is now common to build client-server applications and systems using TCP. The benefit of TCP compared to UDP is that it works reliably over any network. The obvious drawback is that TCP introduces considerably more overhead especially compared to those cases in which underlying network is highly reliable, such as in local area systems.

Middleware

Whereas a DOS attempts to create a specific system for distributed applications, the goal of middle ware is to create system independent interfaces for distributed applications.

Middleware consists of a layer of services added between those of a regular network OSI and the actual applications. These services facilitate the implementation of distributed applications and attempt to hide the heterogeneity of the underlying system architectures (both hardware and software). The principle aim of middleware, namely raising the level of abstraction for distributed programming, is achieved in three ways:

1. Communication mechanisms that are more convenient and less error prone than message passing;
2. Independence from as, network protocol, programming language, etc. and
3. Standard services (such as a naming service).

To make the integration of these various services easier, and to improve transparency and system independence, middleware is usually based on a particular paradigm, or model, for describing distribution and communication. This often manifests itself in a particular programming model such as 'everything is a file', remote procedure call, and distributed objects. Providing such a paradigm automatically provides an abstraction for procedure to follow, and provides direction for how to design and set up the distributed applications. Although some forms of middleware focus on adding support for distributed computing directly into a language (e.g., Erlang), middleware is generally implemented as a set of libraries and tools that enable retrofitting of distributed computing capabilities to existing programming languages.

Such systems typically use a central mechanism of the host language (such as the procedure call and method invocation) and dress remote operations up such that they use the same syntax as that mechanism, resulting for example in remote procedure calls and remote method invocation. Because an important goal of middleware is to hide the heterogeneity of the underlying systems (and in particular of the services offered by the underlying OS), middleware systems often try to offer a complete set of services so that clients do not have to rely on underlying OS services directly. This provides transparency for programmers writing distributed application using the given middleware. Unfortunately this 'everything but the kitchen sink' approach often leads to highly bloated systems. As such, current systems exhibit an unhealthy tendency to include more and more functionality in basic middleware and its extensions, which leads to a jungle of bloated interfaces.

Distributed Shared Memory (DSM)

Because distributed process cannot access each other's memory directly, using shared memory in a distributed system require special mechanisms that emulate the presence of directly accessible shared memory.This is called distributed-shared memory (DSM). The idea behind DSM is that processes on separate computers all have access to the same virtual address space. The memory pages that make up this address space actually reside on separate computers. Whenever a process on one of the computers needs to access a particular page it must find the computer actually hosting that page and request the data from it. Figure shows an example of how a virtual address space might be distributed over various computers.

Message Passing

Message passing in a distributed system is similar to communication using messages in a non-distributed system. The main difference being that the only mechanism available for the passing of messages is network communication. At its core message passing involves two operations send and receive. Although these are very simple operations, there are many variations on the basic model. For example, the communication can be connectionless or connection oriented. Connection oriented communication requires that the sender and receiver first create a connection before send and receive can be used. Communication operations can also be synchronous or asynchronous.

In the first case the operations block until a message has been delivered (or received). In the second case the operations return immediately. Yet another possible variation involves the buffering of communication. In the buffered case, a message will be stored if the receiver is not able to pick it up right away. In the unbuffered case the message will be lost. There are also varying degrees of reliability of the communication. With reliable communication errors are discovered and fixed transparently. This means that the processes can assume that a message that is sent will actually arrive at the destination (as long as the destination process is there to receive it).

Finally it is possible to provide guarantees about the ordering of messages. Thus, for example, a communication system may guarantee that all messages are received in the same order that they are sent, while another system may make no guarantees about the order of arrival of messages. All of these are important issues to consider when dealing with processes that communicate using message passing. Besides these variations in the message passing model, there are also issues involved with communicating between processes on heterogeneous computers.

Different Types of Communication Models

There are numerous ways that communicating processes can be arranged. Following are some common communication models. These models are distinguished from each other by the roles that the communicating processes take on.

Client-server model

The client-server model is the most common and widely used model for communication between processes. As Figure below shows, in this model one process takes on the role of a server, while all other processes take on the roles of clients. The server process provides a service (e.g., a time service, a database service, a banking service, etc.) and the clients are customers of that service. A client sends a request to a server, the request is processed at the server and a reply is returned to the client.

A typical client-server application can be decomposed into three logical parts: the interface part, the application logic part, and the data part. Implementations of the client-server model vary with regards to how the parts are separated over the client and server roles. A thin client implementation will provide a minimal user interface layer, and leave everything else to the server. A fat client implementation, on the other hand, will include all of the user interface and application logic in the client, and only rely on the server to store and provide access to data. Implementations in between will split up the interface or application logic parts over the clients and server in different ways.

Vertical distribution model

An extension of the client-server model, the vertical distribution, or multi-tier, model distributes the traditional server functionality over multiple servers. A client request is sent to the first server. During processing of the request this server will request the services of the next server, who will do the same, until the final server is reached. In this way the various servers become clients of each other.

Each servers is responsible for a different step (or tier) in the fulfillment of the original client request. Splitting up the server functionality in this way is beneficial to a system's scalability as well as its flexibility. Scalability is improved because the processing load on each individual server is reduced, and the whole system can therefore accommodate more users. With regards to flexibility this model allows the internal functionality of each server to be modified as long as the interfaces provided remain the same.

Horizontal distribution model

While vertical distribution focuses on splitting up a server's functionality over multiple computers, horizontal distribution involves replicating a server's functionality over multiple computers. A typical example, as shown in Figure in the next page is a replicated Web server. In this case each server machine contains a complete copy of all hosted Web pages and client requests are passed on to the servers in a round robin fashion. The horizontal distribution model is generally used to improve scalability (by reducing the load on individual servers) and reliability (by providing redundancy). Note that it is also possible to combine the vertical and horizontal distribution models. For example, each of the servers in the vertical decomposition can be horizontally distributed.

Another approach is for each of the replicas in the horizontal distribution model to themselves be vertically distributed.

Peer to peer model (P2P)

Whereas the previous models have all assumed that different processes take on different roles in the communication model, the peer to peer (P2P) model takes the opposite approach and assumes that all processes play the same role, and are therefore peers of each other. In figure below each processes acts as both a client and a server, both sending out requests and processing incoming requests.

Group communication model

The group communication model provides a departure from the point to point style of communication assumed so far. In this model of communication a process can send a single message to a group of other processes. Group communication is often referred to as broadcast (when a single message is sent out to everyone) and multicast (when a single message is sent out to a predefined group of recipients). Group communication can be applied in any of the applied in any of the previously discussed models. It is often used to send requests to a group of replicas, or to send updates to a group of servers containing the same data, It is also used for service discovery (e.g., broadcast a request saying "who offers this service?") as well as event notification (e.g., to tell everyone that the printer is on fire).

Issues involved with implementing and using group communication are similar to those involved with regular point-to-point communication. This includes reliability and ordering. The issues are made more complicated because now there are multiple recipients of a message and different combinations of problems may occur (for example, what if only one of the recipients does not receive a message, should it be multicast out to everyone again, or only to the process that did not receive it?). A widely implemented (but not as widely used) example of group communication is IP multicast. The IP multicast specification has existed for a long time, but it has taken a while for implementations to make their way into the routers and gateways (which is necessary for it to become viable).

Threads in Distributed System

To understand the role of threads in distributed systems, it is important to understand what a process is, and how processes and threads relate. To execute a program, an operating system creates a number of virtual processors, each one for running a different program. To keep track of these virtual processors, the operating system has a process table, containing entries to store CPU register values, memory maps, open files, accounting information, privileges, etc. A process is often defined as a program in execution, that is, a program that is currently being executed on one of the operating system's virtual processors.

An important issue is that the operating system takes great care to ensure that independent processes cannot maliciously or inadvertently affect the correctness of each other's behavior. In other words, the fact that multiple processes may be concurrently sharing the same CPU and other hardware resources is made transparent. Usually, the operating system requires hardware support to enforce this separation. This concurrency transparency comes at a relatively high price. For example, each time a process is created, the operating system must create a complete independent address space.

Allocation can mean initialising memory segments by, for example, zeroing a data segment, copying the associated program into a text segment, and setting up a stack for temporary data. Likewise, switching the CPU between two processes may be relatively expensive as well. Apart from saving the CPU context (which consists of register values, program counter, slack pointer, etc.), the operating system will also have to modify registers of the memory management unit (MMU) and invalidate address translation caches such as in the translation look aside buffer (TLB).

In addition, if the operating system supports more processes than it can simultaneously hold in main memory, it may have to swap processes between main memory and disk before the actual switch can take place. A thread is very similar to a process in the sense that it can also be seen as the execution of a (part of a) program on a virtual processor. However, in contrast to processes, no attempt is made to achieve a high degree of concurrency transparency if this would result in performance degradation. Therefore, a thread system generally maintains only the minimum information to allow a CPU to be shared by several threads.

In particular, a thread context often consists of nothing more than the CPU context, along with some other information for thread management. For example, a thread system may keep track of the fact that a thread is currently blocked on a mutex variable, so as not to select it for execution. Information that is not strictly necessary to manage multiple threads is generally ignored. For this reason, protecting data against inappropriate access by threads within a single process is left entirely to application developers.

There are two important implications of this approach. First, the performance of a multithreaded application need hardly ever be worse than that of its single-threaded counterpart. In fact, in many cases, multithreading leads to a performance gain. Second, because threads are not automatically protected against each other the way processes are, development of multithreaded applications requires some additional intellectual effort. Proper design and keeping things simple, as usual, help a lot. Unfortunately, current practice does not demonstrate that this principle is equally well understood.

There are several benefits to multithreaded processes that have increased the popularity of using thread systems. The most important benefit comes from the fact that in a single-threaded process, whenever a blocking system call is executed, the process as a whole is blocked. To illustrate, consider an application such as spreadsheet program, and assume that a user continuously and interactively wants to change values.

An important property of a spreadsheet program is that it maintains the functional dependencies between different cells, often from different spreadsheets. Therefore, whenever a cell is modified, all dependent cells are automatically updated. When a user changes the value in a single cell, such a modification can trigger a large series of computations. If there is only a single thread of control, computation cannot proceed while the program is waiting for input. Likewise, it is not easy to provide input while dependencies are being calculated. The easy solution is to, have at least two threads of control: one for handling interaction with the user and one for updating the spreadsheet.

Another advantage of multithreading is that it becomes possible to exploit parallelism when executing the program on a multiprocessor system. In that case, each thread is assigned to a different CPU while shared data are stored in shared main memory. When properly designed, such parallelism can be transparent: the process will run equally well on a uniprocessor system, albeit slower. Multithreading for parallelism is becoming increasingly important with the availability of relatively cheap multiprocessor workstations. Such computer systems are typically used for running servers in client-server applications.

Multithreading is also useful in the context of large applications. Such applications are often developed as a collection of co-operating programs, each to be executed by a separate process. This approach is typical for a UNIX environment. Cooperation between programs is implemented by means of inter process communication (IPC) mechanisms. For UNIX systems, these mechanisms typically include (named) pipes, message queues, and shared memory segments. The major drawback of all IPC mechanisms is that communication often requires extensive context switching, shown at three different points in Figure. Because IPC requires kernel intervention, a process will generally first have to switch from user mode to kernel mode, shown as S1 in Figure.

This requires changing the memory map in the MMU, as well as flushing the TLB. Within the kernel, a process context switch takes place (S2 in the figure), after which the other party can be activated by switching from kernel mode to user mode again. The latter switch again requires changing the MMU map and flushing the TLB. Instead of using processes, an application can also be constructed such that different parts are executed by separate threads. Communication between those parts is entirely dealt with by using shared data. Thread switching can sometimes be done entirely in user space, although in other implementations, the kernel is aware of threads and schedules them.

The effect can be a dramatic improvement in performance. Finally, there is also a pure software engineering reason to use threads: many applications are simply easier to structure as a collection of cooperating threads. Think of applications that need to perform several (more or less independent) tasks. For example, in the case of a word processor, separate threads can be used for handling user input, spelling and grammar checking, document layout, index generation, etc.

Threads are generally provided in the form of a thread package. Such a package contains operations to create and destroy threads as well as operations on synchronisation variables such as mutexes and condition variables. There are basically two approaches to implement a thread package. The first approach is to construct a thread library that is executed entirely in user mode. The second approach is to have the kernel be aware of threads and schedule them. A user-level thread library has a number of advantages.

First, it is cheap to create and destroy threads. Because all thread administration is kept in the user's address space, the price of creating a thread is primarily determined by the cost for allocating memory to set up a thread stack. Analogously, destroying a thread mainly involves freeing memory for the stack, which is no longer used. Both operations are cheap.

A second advantage of user-level threads is that switching thread context can often be done in just a few instructions. Basically, only the values of the CPU registers need to be stored and subsequently reloaded with the previously stored values of the thread to which it is being switched. There is no need to change memory maps, flush the TLB, do CPU accounting, and so on. Switching thread context is done when two threads need to synchronise, for example, when entering a section of shared data. However, a major drawback of user-level threads is that invocation of a blocking system call will immediately block the entire process to which the thread belongs, and thus also all the other threads in that process.

Threads are particularly useful to structure large applications into parts that could be logically executed at the same time. In that case, blocking on I/O should not prevent other parts to be executed in the meantime. For such applications, user-level threads are of no help. These problems can be mostly circumvented by implementing threads in the operating system's kernel. Unfortunately, there is a high price to pay: every thread operation (creation, deletion, synchronisation, etc.), will have to be carried out by the kernel, requiring a system call. Switching thread contexts may now become as expensive as switching process contexts. As a result, most of the benefits of using threads instead of processes then disappear. A solution lies in a hybrid form of user-level and kernel-level threads, generally referred to as lightweight processes (LWP).

An LWP runs in the context of a single (heavy-weight) process, and there can be several LWPs per process. In addition to having LWPs, a system also offers a user-level thread package, offering applications the usual operations for creating and destroying threads. In addition, the package provides facilities for thread synchronisation, such as mutexes and condition variables. The important issue is that the thread package is implemented entirely in user space. In other words, all operations on threads are carried out without intervention of the kernel. The thread package can be shared by multiple LWPs. This means that each LWP can be running its own (user-level) thread. Multithreaded applications are constructed by creating threads, and subsequently assigning each thread to an LWP. Assigning a thread to an LWP is normally implicit and hidden from the programmer. The combination of (user-level) threads and LWPs works as follows. The thread package has a single routine to schedule the next thread. When creating an LWP (which is done by means of a system call), the LWP is given its own stack, and is instructed to execute the scheduling routine in search of a thread to execute. If there are several LWPs, then each of them executes the scheduler. The thread table, which is used to keep track of the current set of threads, is thus shared by the LWPs. Protecting this table to guarantee mutually exclusive access, is done by means of mutexes that are implemented entirely in user space. In other words, synchronisation between LWPs does not require any kernel support.

An important property of threads is that they can provide a convenient means of allowing blocking system calls without blocking the entire process in which the thread is running. This property makes threads particularly attractive to use in distributed systems as it makes it much easier to express communication in the form of maintaining multiple logical connections at the same time.

Multithreaded Clients

To establish a high degree of distribution transparency, distributed systems that operate in wide-area networks may need to conceal long interprocess message propagation times. The round-trip delay in a wide-area network can easily be in the order of hundreds of milliseconds, or sometimes even seconds. The usual way to hide communication latencies is to initiate communication and immediately proceed with something else. A typical example where this happens is in Web browsers. In many cases, a Web document consists of an HTML file containing plain text along with a collection of images, icons, etc. To fetch each element of a Web document, the browser has to set up a TCP/IP connection, read the incoming data, and pass it to a display component.

Setting up a connection as well as reading incoming data are inherently blocking operations. When dealing with long-haul communication, we also have the disadvantage that the time for each operation to complete may be relatively long. A Web browser often starts with fetching the HTML page and subsequently displays it. To hide communication latencies as much as possible, some browsers start displaying data while it is still coming in. While the text is made available to the user, including the facilities for scrolling and such, the browser continues with fetching other files that make up the page, such as the images. The latter are displayed as they are brought in.

The user need thus not wait until all the components of the entire page are fetched before the page is made available. In effect, it is seen that the Web browser is doing a number of tasks simultaneously. As it turns out, developing the browser as a multithreaded client simplifies matters considerably. As soon as the main HTML file has been fetched, separate threads can be activated to take care of fetching the other parts. Each thread sets up a separate connection to the server and pulls in the data. Setting up a connection and reading data from the server can be programmed using the standard (blocking) system calls, assuming that a blocking call does not suspend the entire process.

There is another important benefit to using multithreaded Web browsers in which several connections can be opened simultaneously. In the previous example, several connections were set up to the same server. If that server is heavily loaded, or just plain slow, no real performance improvements will be noticed compared to pulling in the files that make up the page strictly one after the other.

Multithreaded Servers

Although there are important benefits to multithreaded clients, the main use of multithreading in distributed systems is found at the server side. Practice shows that multithreading not only simplifies server code considerably, but also makes it much easier to develop servers that exploit parallelism to attain high performance, even on uniprocessor systems. However, now that multiprocessor computers are widely available as general-purpose workstations, multithreading for parallelism is even more useful. To understand the benefits of threads for writing server code, consider the organisation of a file server that occasionally has to block waiting for the disk. The file server normally waits for an incoming request for a file operation, subsequently carries out the request, and then sends back the reply.

Applications

There are two main reasons for using distributed systems and distributed computing. First, the very nature of the application may require the use of a communication network that connects several computers. For example, data is produced in one physical location and it is needed in another location.

Second, there are many cases in which the use of a single computer would be possible in principle, but the use of a distributed system is beneficial for practical reasons. For example, it may be more cost-efficient to obtain the desired level of performance by using a cluster of several low-end computers, in comparison with a single high-end computer. A distributed system can be more reliable than a non-distributed system, as there is no single point of failure. Moreover, a distributed system may be easier to expand and manage than a monolithic uniprocessor system.

Examples of distributed systems and applications of distributed computing include the following:

1. *Telecommunication networks:*
   - Telephone networks and cellular networks.
   - Computer networks such as the Internet.
   - Wireless sensor networks.
   - Routing algorithms.

2. *Network applications:*
   - World wide web and peer-to-peer networks.
   - Massively multiplayer online games and virtual reality communities.
   - Distributed databases and distributed database management systems.
   - Network file systems.
   - Distributed information processing systems such as banking systems and airline reservation systems.
3. *Real-time process control:*
   - Aircraft control systems.
   - Industrial control systems.
4. *Parallel computation:*
   - Scientific computing, including cluster computing and grid computing and various volunteer computing projects.
   - Distributed rendering in computer graphics.

Distributed Systems: Problems and Challenges

Throughout the study of the various aspects of distributed systems, distributed computing system has following challenges and problems:

— Transparency

— Flexibility

— Reliability

— Performance

— Scalability

These challenges can also be seen as the goals of a distributed system. These are briefly described below.

Transparency: Transparency is the concealment from the user and the application programmer of the separation of the components of a distributed system (single image view). Transparency is a strong property that often is difficult to achieve. There are a number of different forms of transparency including the following:

— *Location*: Users are unaware of location of resources

— *Migration*: Resources can migrate without name change Replication: Users are unaware of the existence of multiple copies.

— *Replication*: Users are unaware of the existence of multiple copies.

— *Failure*: Users are unaware of the failure of individual components

— *Concurrency*: Users are unaware of sharing resources with others

— *Parallelism*: Users are unaware of parallel execution of activities

Flexibility: A flexible distributed system can be configured to provide exactly the services tb& a user or programmer needs. A system with this kind of flexibility generally provides a number of key properties. Extensibility allows one to add or replace system components in order to extend or modify system functionality.

— *Openness*: It means that a system provides its services according to standard rules regarding inv0cation syntax and semantics. Openness allows multiple implementations of standard components to be produced. This provides choice and flexibility.

— *Interoperability*: It ensures that systems implementing the same standards (and possibly even those that do not) can interoperate. An important concept with regards to flexibility is the separation of policy and mechanism. A mechanism provides the infrastructure necessary to do something while a policy determines how that something is done. For example, a distributed system may provide secure communication by enabling the encryption of all messages. A system where policy and mechanism is not separated might provide a single hard coded encryption algorithm that is used to encrypt all outgoing messages. A more flexible system, on the other hand, would provide the infrastructure (i.e., the mechanism) needed to call an arbitrary encryption routine when encrypting outgoing messages. In this way the user or programmer is given an opportunity to choose the most appropriate algorithm to use, rather than a built-in system default. Component-based architectures are inherently more flexible than monolithic architectures, which makes them particularly attractive for distributed systems.

Reliability: Although distributed systems provide chance of higher availability by replication, the distributed nature of services means that more components have to work properly for a single service to function. Hence, there are more potential points of failure and when the system architecture does not take explicit measures to increase reliability, there may actually be a degradation of availability. Reliability requires consistency, security, and fault tolerance.

Performance: Any system should strive for maximum performance, but in the case of distributed systems this is a particularly interesting challenge, as it directly conflicts with some other desirable properties. In particular, security, transparency, the ability for migration, and reliability can easily be detrimental to performance.

Scalability: Scalability is important in distributed systems, and in particular in wide area distributed systems, and those expected to experience large growth. According to Neumann a system is scalable if it can handle the addition of users and resources without suffering a noticeable loss of performance or increase in administrative complexity. Adding users and resources causes a system to grow. This growth has three dimensions:

— *Size*: A distributed system can grow with regards to the number of users or resources (e.g., computers) that it supports. As the number of users grows the system may become overloaded. Likewise as the number of resources managed by the system grows the administration that the system has to perform may become too overwhelming for it.

— *Geography*: A distributed system can grow with regards to geography or the distance between nodes. An increased distance generally results in greater communication delays and the potential for communication failure. Another aspect of geographic scale is the clustering of users in a particular area. While the whole system may have enough resources to handle all users, because they are all concentrated in a single area, the resources available there may not be sufficient to handle the load.

— *Administration*: As a distributed system grows, its various components (users, resource, nodes. networks, etc.) will start to cross administrative domains. This means that the number- of organisations or individuals that exert administrative control over the system will grow. k a system that scales poorly with regards to administrative growth this can lead to problems of resource usage, reimbursement, security,. etc. In short, an administrative mess. A claim often made for newly introduced distributed systems is that they are scalable. These claims of scalability are often (unintentionally) unfounded because they focus on very limited aspects of scalability. Although this is a valid claim it says nothing about the scalability with regards to users, or geographic distribution, for example.

Another problem with claims of scalability is that many of the techniques used to improve scalability (such as replication), introduce new problems which are often fixed using non-scalable solutions (e.g., the solutions for keeping replicated data consistent may be inherently nonscalable). Note also that, although a scalable system requires that growth does not affect performance adversely, the mechanisms to make the system scalable may have adverse effects the overall system performance (e.g., the performance of the system when deployed in a small scale (i.e., when scalability is not as important) may be less than optimal due to the overhead of the scalability mechanisms). The key approach to designing and building a scalable system is decentralisation. Generally this requires avoiding any form of centralisation, as this can cause performance bottlenecks. In particular a scalable distributed system must avoid centralising:

— Components (e.g., avoid having a single server),
— Tables (e.g., avoid having a single centralised directory of names), and
— Algorithms (e.g., avoid algorithms based on complete information).
— When designing algorithms for distributed systems the following design rules can help avoid centralisation:
— Do not require any machine to hold complete system state.
— Allow nodes to make decisions based on local info.
— Algorithms must survive failure of nodes.
— No assumption of a global clock.

Other, more specific, approaches to avoiding centralisation and improving scalability include: Hiding (or masking) communication delays introduced by wide area networks; Distributing data over various machines to reduce the load placed on any single machine; and Creating replicas of data and services to reduce the load on any single machine. Besides spreading overall system load out over multiple machines, distribution and replication also help to bring data closer to the users, thus improving geographic scalability. Furthermore, by allowing distribution of data, the management and responsibility of data can be kept within any particular administrative domain, thus helping to improve administrative scalability as well.

References

Attiya, Hagit and Welch, Jennifer (2004), *Distributed Computing: Fundamentals, Simulations, and Advanced Topics*, Wiley-Interscience.

Cachin, C.; Guerraoui, R.; Rodrigues, L. S. (2011). *Introduction to Reliable and Secure Distributed Programming*.

Faber, Jim (1998), *Java Distributed Computing*, O'Reilly: Java Distributed Computing by Jim Faber.

Garg, Vijay K. (2002), *Elements of Distributed Computing*, Wiley-IEEE Press.

Tel, Gerard (1994), *Introduction to Distributed Algorithms*, Cambridge University Press.

5

Networking

A computer network is a collection of computers and other hardware interconnected by communication channels that allow sharing of resources and information. Where at least one process in one device is able to send/ receive data to/from at least one process residing in a remote device, then the two devices are said to be in a network. A network is a group of devices connected to each other. Networks may be classified into a wide variety of characteristics, such as the medium used to transport the data, communications protocol used, scale, topology, benefit, and organizational scope.

Communications protocols define the rules and data formats for exchanging information in a computer network, and provide the basis for network programming. Well-known communications protocols include two Ethernet, a hardware and link layer standard that is ubiquitous in local area networks, and the Internet protocol suite, which defines a set of protocols for internetworking, i.e. for data communication between multiple networks, as well as host-to-host data transfer, and application-specific data transmission formats.Computer networking is sometimes considered a sub-discipline of electrical engineering, telecommunications, computer science, information technology or computer engineering, since it relies upon the theoretical and practical application of these disciplines.

The basis reasons why computers are networked are cost reduction by sharing hardware & software resources. High reliability by having multiple sources of supply. Cost reduction by downsizing to microcomputer based networks instead of using mainframes.Greater flexibility because of possibility to connect devices from various vendors.Increase productivity by making it easier to share data amongst users Take for example a typical office scenario where a number of users in a small business require access to common information. As long as all user computers are connected via a network, they can share their files, exchange mail, schedule meetings, send faxes and print documents all from any point of the network. It would not be necessary for users to transfer files via electronic mail or floppy disk, rather, each user could access all the information they require, thus leading to less wasted time and hence greater productivity.

Evolution of Networks

Computer Networking has evolved under several different models to respond to different needs and according to the way they process data. These models are Centralised, Distributed, and Collaborative.

Centralised Computing

In early computer systems around 1950's,were large, difficult to manage and expensive. These central computers were called as mainframes and they were used to store, process and arrange data. Jobs were entered into the system by reading commands from card decks. The computer would execute one job at a time and generate a printout when the job was complete. Terminals, which enabled users to interact with the centralised computer, were a much later development.

In the computing environment of the mainframe world, all processing and data storage are centralised in the mainframe computer. Terminals are simple devices that display characters on screens and accept typed input. Networks developed when it became necessary for the mainframe computers to share information and services.

Advantages of centralised computing

— Ease of back up

— Security.

— Low cost.

Disadvantages of centralised computing:

— Slow network access.

— Fewer options.

Distributed Computing

As personal computers were introduced to organisations, a new model of distributed computing emerged. Instead of concentrating computing in a central device, PCs made it possible to give each worker an individual computer. Each PC can process and store data independently.

Under the distributed computing model, networking has evolved to enable the many distributed computers to exchange and share resources and services.

Advantages of distributed computing

— Quick access.

— Multiple uses.

Disadvantages of distributed computing

— Virus susceptibility.

— Backup difficulty.

— File synchronisation.

Collaborative Computing

Also called Cooperative computing, collaborative computing enables computers in a distributed computing environment to share processing power in addition to data, resources and services. In collaborative computing environment, computers might "borrow" processing power by running programs on other computers on the network, or processes might be designed so they will run on two or more computers. Obviously, collaborative computing cannot take place without a network to enable the various computers to communicate.

Advantages of collaborative computing

— Extremely fast.

— Multiple uses.

Disadvantages of distributed computing

— Susceptible to viruses.

— Difficult to back up.

— Difficult file synchronisation.

Network Architecture

Client/server Architecture

Client/Server architecture is one in which the client (personal computer or workstation) is the requesting machine and the server is the supplying machine, both of which are connected via a local area network (LAN) or wide area network (WAN). Since the early 1990s, client/server has been the buszword for building applications on LANs in contrast to centralised minis and mainframes with dedicated terminals.

The client contains the user interface and may perform some or all of the application processing. Servers can be high-speed microcomputers, minicomputers or even mainframes. A database server maintains the databases and processes requests from the client to extract data from or update the database. An application server provides additional business processing for the clients.

However, client/server architecture means more than dedicated servers. Simply downloading files from or sharing programs and databases on a server is not true client/server either. True client/ server implies that the application was originally designed to run on a network and that the network infrastructure provides the same quality of service as traditional mini and mainframe information systems.

The network operating system (NOS) together with the database management system (DBMS) and transaction monitor (TP monitor) are responsible for integrity and security.

Non-client/Server

In non-client/server architecture, the server is nothing more than a remote disk drive. The uzser's machine does all the processing. If many users routinely perform lengthy searches, this can bog down the network, because each client has to pass the entire database over the net. At 1,000 bytes per record, a 10,000 record database requires 10MB of data be transmitted.

Two-tier Client/Server

Two-tier client/server is really the foundation of client/server. The database processing is done in the server. An SQL request is generated in the client and transmitted to the server. The DBMS searches locally and returns only matching records. If 50 records met the criteria, only 50K would be transmitted. This reduces traffic in the LAN.

Three-Tier Client/Server

Many applications lend themselves to centralised processing. If they contain proprietary algorithms, security is improved. Upgrading is also simpler. Sometimes, programs are just too demanding to be placed into every client PC. In three-tier client/ server, application processing is performed in one or more servers.

Peer – to – Peer Architecture

A type of network in which each workstation has equivalent capabilities and responsibilities. This differs from client/server architectures, in which some computers are dedicated to serving the others. Peer-to-peer networks are generally simpler and less expensive, but they usually do not offer the same performance under heavy loads.

Network Basics

In many organisations today, computers are connected together by cable or by phone line in a network. In a network, a computer can be attached to computers in adjoining offices, adjoining buildings, and to computers around the globe. If you are connected to a network, you can open and save files on other computers or use resources attached to those computers, like printers and CD-ROM drive. Of course, you must have been given the right to do this.

To access other computers on your network, open the Network Neighbourhood object on the desktop. If your network is large, you can view the computers in your immediate workgroup. A workgroup is a collection of computers and computer users that share similar resources and belong to the same group. Whereas, domain that is similar to a workgroup is much larger in scale. Each domain has its own computers and workgroups and is managed independently of other domain.

Setup and Configuration

Assuming that the network administrator has already installed the network and established your computer as a workstation, you should be able to identify your computer network settings, like - the network name, the workgroup it belongs to, and how folders, drives, and other resources are shared on your computer. To find out the network settings, double-click the Network object from the Control Panel window. In the Network window that is opened up, you will find three tabs - Configuration, Identification, and Access Control.

Configuration will display the network components that are installed, or removed, or properties be viewed. These are :

Client Software - It enables you to use files and printers shared on other network computers.

An Adapter - It is a hardware device that physically connects your computer to the network.

A Protocol - It is the language a computer uses to communicate over a network. All computers connected to the same network must use the same protocol to communicate with each other.

In the same window, the Primary Logon is the client that is used to validate your user name and password, process any login script, and perform other startup tasks.

Identification tab will open up the dialog box that displays the information to identify your computer on the network. In this, the Computer name is the name of your computer that other people see when they browse the network. The Workgroup is the name of the local network that your network belongs to. The Computer Description text box provides the place to enter the description for your computer. This information will be viewed by other users when they browse the network.

Access Control tab provides you the option to determine how the files on your computer and the resources attached to it are shared over the network. The two options are :

Share-level access control is used to specify the passwords that other network users must type to access the resources you share on your computer.

User-level access control enables you to validate or authenticate users before they can access your computer. This works only if your network has a special security server.

Logging onto the Network

The Network Administrator assigns the login name and the password to every person who will use the network. If you are already working on Windows 95, to log on to the network, follow the given instructions :

1. Click the Start button, and then click Shut Down.

2. Click Close All Programs And Log On As A Different User.
3. Click Yes.
4. In the Enter Network Password box, type your user name and password.

Mapping Network Drives

In order to access folders on network drives from some applications, a drive letter must be assigned to these folders. These drives are not physical drives but the mapping is done for logical drives.

To map a drive letter to a network computer or folder, follow the given steps :

1. In My Computer or Network Neighborhood, locate the folder you want to map.
2. Right-click the folder to open its context menu.
3. In the Drive box, the next recommended drive is displayed. You can click on the down-arrow button to choose a different drive letter. In any case, it should not be the drive letter that is already in use.
4. Type a path to the computer and folder in the Path field.
5. Click OK.

To disconnect a mapped drive to make it free for other mappings, follow the given steps :

1. Open the My Computer window.
2. Right-click the mapped object you want to disconnect.
3. From its context menu, select Disconnect to remove the mapping.

Network Access

Network is defined as interconnection of two or more computers such that they can resources and information. These computers can be linked together using a wide variety of cables or telephone lines or through satellites.

There are many types of computer networks, including:

— *Local Area Networks (LANs)*: The computers are geographically close together (that is, in the same building).

— *Metropolitan Area Network (MAN)*: The Computer network designed for a town or city.

— *Wide Area Networks (WANs)*: The computers are farther apart and are connected by telephone lines or radio waves. In addition to these types, the following characteristics are also used to categorise different types of networks:

— *Topology*: The geometric arrangement of a computer system. Common topologies include a bus, star, and ring.

— *Protocol* : The protocol defines a common set of rules and signals that computers on the network use to communicate. One of the most popular protocols for LANs is called Ethernet. Another popular LAN protocol for PCs is the IBM token-ring network.

— *Architecture* : Networks can be broadly classified as using either a peer-to-peer or client/server architecture. Computers on a network are sometimes called nodes. Computers and devices that allocate resources for a network are called servers.

E-Mail

In its simplest form, e-mail is an electronic message sent from one computer to another. You can send or receive personal and business-related messages with attachments, such as pictures or formatted documents. You can even send computer programs.

How e-mail works

It can take days to send a letter across the country and weeks to go around the world. To save time and money, more and more people are using electronic mail. It's fast, easy and much cheaper than the post office.

Let's say you have a small business with sales reps working around the country. How do you keep in touch without running up a huge phone bill? Or what about keeping in touch with far-flung family members? E-mail is the way to go. Its no wonder e-mail has become the most popular service on the Internet.

Just as a letter makes stops at different postal stations along its way, e-mail passes from one computer, known as a mail server to another as it travels over the Internet. Once it arrives at the destination mail server, it is stored in an electronic mailbox until the recipient retrieves it. This whole process can take seconds, allowing you to quickly communicate with people around the world at any time of the day or night.

To receive e-mail, you must have an account on a mail server. This is similar to having an address where you receive letters. One advantage over regular mail is that you can retrieve your email from a remote location. Once you contact your mail server, you can download your messages.

To send e-mail, you need a connection to the Internet and access to a mail server, which can forward your mail. The standard protocol used for sending e-mail on the Internet is called SMTP, for Simple Mail Transfer Protocol. It works in conjunction with POP servers. POP stands for Post Office Protocol. When you send an e-mail message, your computer sends it to an SMTP server. The server looks at the e-mail address (like the address on an envelope), and then forwards it to the recipient's mail server. When the message is received at the destination mail server, it is stored until the addressee retrieves it. You can send e-mail to anyone who has an e-mail address, anywhere in the world. Remember, almost all Internet service providers (ISPs) and all major online services offer an e-mail address with every account.

Sending An E-Mail Message

Step - 1

— Open your e-mail program and launch a new message window by clicking on the appropriate icon.

Step - 2

— In the TO box, type in the name of the recipient. It should take this form:

— recipient@domain.com

— Make sure you enter the address correctly or the message will return to you.

— You can send a message to more than one person by entering multiple addresses. Just put a semi-colon (;) between each address.

— Your return address is automatically sent to the recipient.

Step - 3

— Type in the subject of the e-mail.

Step - 4

— Write your message in the message window. You can also copy text from a word processing program and paste it into the window.

Step- 5

— Click on the Send icon or select Send from the File menu.

Understanding e-mail addresses

Internet e-mail addresses typically have two main parts: professor@raiuniversity.edu First there is the user name (professor), which refers to the recipient's mailbox. Then there's an axon sign (@). Next comes the host name (rai university), also called the domain name. This refers to the mail server, the computer where the recipient has an electronic mailbox. It's usually the name of a company or organisation. Finally, there's a dot (.) followed by three letters (com) that indicate the type of domain.

An address ending with .com typically means that the host is a business, commercial enterprise, or an online service like America Online. Most companies use this extension. A host name ending with .edu usually means that the host is a university or educational facility. .org indicates the host is a noncommercial organisation.

Other extensions you might encounter are .mil for military and .net for network. The latter tends to be reserved for organisations such as Internet service providers. By the way, there are plans to add seven additional top-level domains, such as .web and .nom, but when this will happen is anyone's guess. For e-mail addresses outside of the United States, there is often a [DOT] followed by two letters representing the country. For instance, .in indicates India, .de indicates Germany and .nz indicates New Zealand.

LAN

LAN is a computer network that spans a relatively small area. Most LANs are confined to a single building or group of buildings. However, one LAN can be connected to other LANs over any distance via telephone lines and radio waves. A system of LANs connected in this way is called a wide-area network (WAN).

Most LANs connect workstations and personal computers. Each node (individual computer) in a LAN has its own CPU with which it executes programs, but it is also able to access data and devices anywhere on the LAN. This means that many users can share expensive devices, such as laser printers, as well as data. Users can also use the LAN to communicate with each other, by sending e-mail or engaging in chat sessions. There are many different types of LANs Ethernets being the most common for PCs..

The following characteristics differentiate one LAN from another:

Topology : The geometric arrangement of devices on the network. For example, devices can be arranged in a ring or in a straight line.

Protocols : The rules and encoding specifications for sending data. The protocols also determine whether the network uses a peer-to-peer or client/server architecture.

Media : Devices can be connected by twisted-pair wire, coaxial cables, or fiber optic cables. Some networks do without connecting media altogether, communicating instead via radio waves.

LANs are capable of transmitting data at very fast rates, much faster than data can be transmitted over a telephone line; but the distances are limited, and there is also a limit on the number of computers that can be attached to a single LAN.

MAN

MAN stands for Metropolitan Area Network, it is a data network designed for a town or city. In terms of geographic breadth, MANs are larger than local-area networks (LANs), but smaller than wide-area networks (WANs). MANs are usually characterised by very high-speed connections using fiber optical cable or other digital media.

WAN

WAN is a computer network that spans a relatively large geographical area. Typically, a WAN consists of two or more local-area networks (LANs).

Computers connected to a wide-area network are often connected through public networks, such as the telephone system. They can also be connected through leased lines or satellites. The largest WAN in existence is the Internet.

Client/Server Architecture

Client/Server Architecture is one in which the client (personal computer or workstation) is the requesting machine and the server is the supplying machine, both of which are connected via a local area network (LAN) or wide area network (WAN). Since the early 1990s, client/server has been the buzzword for building applications on LANs in contrast to centralised minis and mainframes with dedicated terminals. The network operating system (NOS) together with the database management system (DBMS) and transaction monitor (TP monitor) are responsible for integrity and security. Some of these products have gone through many client/server versions by now and have finally reached industrial strength.

The client contains the user interface and may perform some or all of the application processing. Servers can be high-speed microcomputers, minicomputers or even mainframes. A database server maintains the databases and processes requests from the client to extract data from or update the database. An application server provides additional business processing for the clients.

The term client/server is sometimes used to contrast a peer-topeer network, in which any client can also act as a server. In that case, client/server means nothing more than having a dedicated server.

However, client/server architecture means more than dedicated servers. Simply downloading files from or sharing programs and databases on a server is not true client/server either. True client/server implies that the application was originally designed to run on a network and that the network infrastructure provides the same quality of service as traditional mini and mainframe information systems.

Non-Client/Server

In non-client/server architecture, the server is nothing more than a remote disk drive. The user's machine does all the processing. If many users routinely perform lengthy searches, this can bog down the network, because each client has to pass the entire database over the net. At 1,000 bytes per record, a 10,000 record database requires 10MB of data be transmitted.

Two-tier Client/Server

Two-tier client/server is really the foundation of client/server. The database processing is done in the server. An SQL request is generated in the client and transmitted to the server. The DBMS searches locally and returns only matching records. If 50 records met the criteria, only 50K would be transmitted. This reduces traffic in the LAN.

Three-tier Client/Server

Many applications lend themselves to centralised processing. If they contain proprietary algorithms, security is improved. Upgrading is also simpler. Sometimes, programs are just too demanding to be placed into every client

PC. In three-tier client/ server, application processing is performed in one or more servers.

Web Browsers

A web browser is the software program you use to access the World Wide Web, the graphical portion of the Internet. The first browser, called NCSA Mosaic, was developed at the National Center for Supercomputing Applications in the early '90s. The easy-to-use point-and-click interface helped popularise the Web, although few then could imagine the explosive growth that would soon occur.

Although many different browsers are available, Microsoft Internet Explorer and Netscape Navigator are the two most popular ones. Netscape and Microsoft have put so much money into their browsers that the competition can't keep up. The pitched battle between the two companies to dominate the market has lead to continual improvements to the software. Version 4.0 and later releases of either browser are excellent choices. You can download Explorer and Navigator for free from each company's website. If you have one browser already, you can test out the other. Also note that there are slight differences between the Windows and MacIntosh versions.

Browsers come loaded with all sorts of handy features. Fortunately, you can learn the basics in just a few minutes, then take the time to explore the advanced functions. Both Explorer and Navigator have more similarities than differences, so we'll primarily cover those. For the most up-todate information about the browsers, and a complete tutorial, check the online handbook under the Help menu or go to the websites of the respective software companies.

Browser anatomy

When you first launch your web browser, usually by doubleclicking on the icon on your desktop, a predefined web page, your home page, will appear. With Netscape Navigator for instance, you will be taken to Netscape's NetCenter.

Toolbar

The row of buttons at the top of your web browser, known as the toolbar, helps you travel through the web of possibilities, even keeping track of where you've been. Since the toolbars for Navigator and Explorer differ slightly, we'll first describe what the buttons in common do:

The Back button returns you the previous page you've visited. Use the Forward button to return to the page you just came from.

Home takes you to whichever home page you've chosen. (If you haven't selected one, it will return you to the default home page, usually the Microsoft or Netscape website.) Reload or Refresh does just that, loads the web page again. Why would you want to do this? Sometimes all of the elements of a web page haven't loaded the first time, because the file transfer was interrupted. Also when you download a web page, the data is cached (pronounced cashed), meaning it is stored temporarily on your computer. The next time you want that page, instead of requesting the file from the web server, your web browser just accesses it from the cache. But if a web page is updated frequently, as may be the case with news, sports scores or financial data, you won't get the most current information. By reloading the page, this timely data is updated.

Print lets you make a hard copy of the current document loaded in your browser.

The Stop button stops the browser from loading the current page.

Search connects to pages on the Microsoft or Netscape websites that list a number of Internet directories and search tools.

Bookmarks or Favorites is where you can record the addresses of website you want to revisit. Once you add a URL to your list, you can return to that web page simply by clicking on the link in your list, instead of retyping the entire address.

Location bar

Just under the toolbar, you will see a box labeled "Location," "Go To," or "Address." This is where you type in the address of a website you want to visit. After you enter it, press the Return or Enter key to access the site.

By clicking the small triangle to the right of the Location box, you will get a drop down list of the most recent websites you have visited. To revisit a site, just click on the address.

Menu bar

Located along the top of the browser window, the menu bar offers a selection of things you can do with a web page, such as saving it to your hard drive or increasing the size of the text on a page. Many of the choices are the same as the buttons on the toolbar below, so don't try to learn everything now. Click once on a word to access the drop-down menu, then click on the appropriate selection you want to make.

Access indicator

Both Navigator and Explorer have a small picture in the upper right hand corner of the browser. When this image is animated, it means that your browser software, known as a client, is accessing data from a remote computer, called a server. The server can be located across town or on another continent. Your browser downloads these remote files to your computer, then displays them on your screen. The speed of this process depends a number of factors: your modem speed, your Internet service provider's modem speed, the size of the files you are downloading, how busy the server is and the traffic on the Internet.

The typical characteristics of a WAN are :

— Can be as large as worldwide.

— Usually much slower than LAN speed.

— Highest possible error rate compared to LAN & MAN.

— Expensive equipment.

Network Application

Sharing Resources

To share any resource on your computer, File and Printer Sharing must first be enabled. This is under Control Panel->Networks->File and Printer Sharing.

Once this is enabled, passwords may be assigned to each resource that is made available. Users cannot use that resource unless they know the password.

To share a resource such as a directory where sales reports are kept, start Windows Explorer and right mouse click the directory that is to be shared. In the diagram below, this has been done on the sub-directory temp on drive c:. Clicking on the Sharing property brings up the following dialog box.

This allows the user to specify a password and name to the resource. In Windows format, the name of the resource is then known on the workgroup as

\\computername\resourcename

For instance, if the computer name was sue, and the resource was specified as temp, then the resource is known as \\sue\temp This is known as the Uniform Naming Convention [UNC] for the resource.

Network Resources

Hardware Resources of Networks

Network interface card

It is also known as Network adapter. It is a printed circuit board that plugs into both the clients (personal computers or workstations) and servers and controls the exchange of data between them. The network adapter provides services at the data link level of the network, which is also known as the access method (OSI layers 1 and 2).

The most common network adapters are Ethernet and Token Ring. Sometimes, the Ethernet adapter is built into the motherboard. LocalTalk, which provides the data link services of Apple's AppleTalk network, is built into all Macintoshes. A transmission medium, such as twisted pair, coax or fiber optic, interconnects all the adapters in the network. A network adapter is also called a NIC, or network interface card.

Repeater

A communications device that amplifies or regenerates the data signal in order to extend the transmission distance. Available for both analog and digital signals, it is used extensively in long distance transmission. It is also used to tie two LANs of the same type together. Repeaters work at layer 1 of the OSI model. The term may also refer to a multiport repeater, which is a hub in a 10BaseT network.

Bridge

Bridge is a device that connects two LAN segments together, which may be of similar or dissimilar types, such as Ethernet and Token Ring. A bridge is inserted into a network to segment it and keep traffic contained within the segments to improve performance.

Bridges learn from experience and build and maintain address tables of the nodes on the network. By monitoring which station acknowledged receipt of the address, they learn which nodes belong to the segment.

Bridges work at the data link layer (OSI layer 2), whereas routers work at the network layer (layer 3). Bridges are protocol independent; routers are protocol dependent. Bridges are faster than routers because they do not have to read the protocol to glean routing information.

Bridges with more than two ports (multiport bridges) perform a switching function. Today's LAN switches are really multiport bridges that can switch at full wire speed.

Transparent bridge

A common type of network bridge, in which the host stations are unaware of their existence in the network. A transparent bridge learns which node is connected to which port through the experience of examining which node responds to each new station address that is transmitted. Ethernet uses this type of bridge, also called an adaptive bridge.

Hub

A central connecting device in a network that joins communications lines together in a star configuration. Passive hubs are just connecting units that add nothing to the data passing through them. Active hubs, also sometimes called multiport repeaters, regenerate the data bits in order to maintain a strong signal, and intelligent hubs provide added functionality. Hubs are mandatory in 10BaseT twisted pair Ethernet as well as Token Ring networks. They are also used to replace the daisy chain cabling in 10Base5 and 10Base2 coaxial Ethernets in order to improve network management.

In Token Rings, the hub is called a MAU (Multi-station Access Unit). Multiple media hubs interconnect different types of Ethernets (twisted pair, coax and optical fiber) and can bridge between Ethernet, Token Ring, FDDI and ATM topologies. Switching hubs provide Ethernet and ATM switching. Hubs have become very intelligent, modular and customisable, allowing for the insertion of bridging, routing and switching modules all within the same unit. A hub can even host a CPU board and network operating system, turning the hub into a file server or some type of network control processor that performs LAN emulation or other complex function as networks grow.

Router

Router is a device that forwards data packets from one local area network (LAN) or wide area network (WAN) to another. Based on routing tables and routing protocols, routers read the network address in each transmitted frame and make a decision on how to send it based on the most expedient route (traffic load, line costs, speed, bad lines, etc.). Routers work at layer 3 in the protocol stack, whereas bridges and switches work at the layer Routers are used to segment LANs in order to balance traffic within workgroups and to filter traffic for security purposes and policy management. Routers are also used at the edge of the network to connect remote offices. Multiprotocol routers support several protocols such as IP, IPX, AppleTalk and DECnet.

Routers can only route a message that is transmitted by a routable protocol such as IP or IPX. Messages in non-routable protocols, such as NetBIOS and LAT, cannot be routed, but they can be transferred from LAN to LAN via a bridge. Because routers have to inspect the network address in the protocol, they do more processing and add more overhead than a bridge or switch, which both work at the data link (MAC) layer. Most routers are specialised computers that are optimised for communications; however, router functions can also be implemented by adding routing software to a file server. NetWare, for example, includes routing software. The NetWare operating system can route from one subnetwork to another if each one is connected to its own network adapter (NIC) in the server.

Routers serve as an internet backbone, interconnecting all networks in the enterprise. This architecture strings several routers together via a high-speed LAN topology such as Fast Ethernet or Gigabit Ethernet. Routers are also the backbone of the Internet, which spans the planet.

Another approach within an enterprise is the collapsed backbone, which uses a single router with a high-speed backplane to connect the subnets, making network management simpler and improving performance.

In older Novell terminology, a router is a network-layer bridge. Routers also used to be called gateways.

Gateway

Gateway is a computer that performs protocol conversion between different types of networks or applications. For example, a gateway can convert a TCP/IP packet to a NetWare IPX packet and vice versa, or from AppleTalk to DECnet, from SNA to AppleTalk and so on.

Gateways function at layer 4 and above in the OSI model. They perform complete conversions from one protocol to another rather than simply support one protocol from within another, such as IP tunneling. Sometimes routers can implement gateway functions.

An electronic mail, or messaging, gateway converts messages between two different messaging protocols. Gateway is a computer that acts as a go-between two or more networks that use the same protocols. In this case, the gateway functions as an entry/exit point to the network. Transport protocol conversion may not be required, but some form of processing is typically performed.

Modem

MODEM is an acronym for modulator-demodulator. A modem is a device that enables a computer to transmit data over telephone lines. Computer information is stored digitally, whereas information transmitted over telephone lines is transmitted in the form of analog waves. A modem converts between these two forms.

Fortunately, there is one standard interface for connecting external modems to computers called RS-232. Consequently, any external modem can be attached to any computer that has an RS-232 port, which almost all personal computers have. There are also modems that come as an expansion board that you can insert into a vacant expansion slot. These are sometimes called onboard or internal modems.

While the modem interfaces are standardised, a number of different protocols for formatting data to be transmitted over telephone lines exist. Some, like CCITT V.34, are official standards, while others have been developed by private companies. Most modems have built-in support for the more common protocols — at slow data transmission speeds at least, most modems can communicate with each other. At high transmission speeds, however, the protocols are less standardised. Aside from the transmission protocols that they support, the following characteristics distinguish one modem from another:

bps : How fast the modem can transmit and receive data. At slow rates, modems are measured in terms of baud rates. The slowest rate is 300 baud (about 25 cps). At higher speeds, modems are measured in terms of bits per second (bps). The fastest modems run at 57,600 bps, although they can achieve even higher data transfer rates by compressing the data. Obviously, the faster the transmission rate, the faster you can send and receive data. Note, however, that you cannot receive data any faster than it is being sent.

If, for example, the device sending data to your computer is sending it at 2,400 bps, you must receive it at 2,400 bps. It does not always pay, therefore, to have a very fast modem. In addition, some telephone lines are unable to transmit data reliably at very high rates.

voice/data: Many modems support a switch to change between voice and data modes. In data mode, the modem acts like a regular modem. In voice mode, the modem acts like a regular telephone. Modems that support a voice/data switch have a built-in loudspeaker and microphone for voice communication.

auto-answer: An auto-answer modem enables your computer to receive calls in your absence. This is only necessary if you are offering some type of computer service that people can call in to use.

data compression : Some modems perform data compression, which enables them to send data at faster rates. However, the modem at the receiving end must be able to decompress the data using the same compression technique.

flash memory : Some modems come with flash memory rather than conventional ROM, which means that the communications protocols can be easily updated if necessary.

Fax capability: Most modern modems are fax modems, which means that they can send and receive faxes.

Firewall

Firewall is a method for keeping a network secure. It can be implemented in a single router that filters out unwanted packets, or it may use a combination of technologies in routers and hosts. Firewalls are widely used to give users access to the Internet in a secure fashion as well as to separate a company's public Web server from its internal network. They are also used to keep internal network segments secure. For example, a research or accounting subnet might be vulnerable to snooping from within.

Following are the types of techniques used individually or in combination to provide firewall protection.

Packet Filter

Blocks traffic based on IP address and/or port numbers. Also known as a "screening router."

Proxy Server

Serves as a relay between two networks, breaking the connection between the two. Also typically caches Web pages.

Network Address Translation (NAT)

Hides the IP addresses of client stations in an internal network by presenting one IP address to the outside world. Performs the translation back and forth.

Stateful inspection

Tracks the transaction in order to verify that the destination of an inbound packet matches the source of a previous outbound request. Generally can examine multiple layers of the protocol stack, including the data, if required, so blocking can be made at any layer or depth.

Switch

In networks, Switch is a device that filters and forwards packets between LAN segments. Switches operate at the data link layer (layer 2) of the OSI Reference Model and therefore support any packet protocol. LANs that use switches to join segments are called switched LANs or, in the case of Ethernet networks, switched Ethernet LANs.

The other defination of switch is : A small lever or button. The switches on the back of printers and on expansion boards are called DIP switches. A switch that has just two positions is called a toggle switch.

Networking Software

Networking support is typically provided by two software components:

— High-Level Networking Software.

— Network Driver Software.

High-level Networking Software

High-Level Networking Software provides end-user-oriented functions that are associated with the Application layer through the Network layer of the OSI model. This is the software that the end user perceives. Some types of high-level networking software subsystems, especially in the personal computer environment, are called network operating systems.

Network Driver Software

Network Driver Software provides an interface between the high-level networking software and the particular network interface card (NIC) that is being used for physical LAN communication. Like the NIC itself, the driver software is generally transparent to the end user.

Commonly Used High Level Networking Software Systems

PPP

Short for Point-to-Point Protocol, a method of connecting a computer to the Internet. PPP is more stable than the older SLIP protocol and provides error checking features. It is a data link protocol that provides dial-up access over serial lines. It can run on any full-duplex link from POTS to ISDN to high-speed lines (T1, T3, etc.). Developed by the Internet Engineering Task Force in 1991, it has become popular for Internet access as well as a method for carrying higher level protocols. Over ISDN, PPP uses one 64 Kbps B channel for transmission. The Multilink PPP protocol (MP, MPPP or MLPPP) bridges two or more B channels for higher-speed operation. For example, using ISDN's Basic Rate service (BRI), you can obtain 128 Kbps with Multilink PPP.

PPP encapsulates protocols in specialised Network Control Protocol packets; for example, IPCP (IP over PPP) and IPXCP (IPX over PPP). It can be used to replace a network adapter driver, allowing remote users to log on to the network as if they were inhouse. PPP can hang up and redial on a low-quality call. PPP also provides password protection using the Password Authentication Protocol (PAP) and the more rigorous Challenge Handshake Authentication Protocol (CHAP).

SLIP

(Serial Line IP) A data link protocol for dial-up access to TCP/ IP networks. It is commonly used to gain access to the Internet as well as to provide dial-up access between two LANs. SLIP transmits IP packets over any serial link (dial up or private lines)

FTP

(File Transfer Protocol) A protocol used to transfer files over a TCP/IP network (Internet, UNIX, etc.). It includes functions to log onto the network, list directories and copy files. It can also convert between the ASCII and EBCDIC character codes. FTP operations can be performed by typing commands at a command prompt or via an FTP utility running under a graphical interface such as Windows. FTP transfers can also be initiated from within a Web browser by entering the URL preceded with ftp://.

Unlike e-mail programs in which graphics and program files have to be "attached," FTP is designed to handle binary files directly and does not add the overhead of encoding and decoding the data.

TELNET

A terminal emulation protocol commonly used on the Internet and TCP/IP-based networks. It allows a user at a terminal or computer to log onto a remote device and run a program. Telnet was originally developed for ARPAnet and is an inherent part of the TCP/IP communications protocol. Although most computers on the Internet require users to have an established account and password, there are many that allow public access to certain programs, typically, search utilities, such as Archie or WAIS.

SMTP

(Simple Mail Transfer Protocol) The standard e-mail protocol on the Internet. It is a TCP/IP protocol that defines the message format and the message transfer agent (MTA), which stores and forwards the mail. SMTP was originally designed for only ASCII text, but MIME and other encoding methods enable program and multimedia files to be attached to e-mail messages.

SMTP servers route SMTP messages throughout the Internet to a mail server, such as POP3 or IMAP4, which provides a message store for incoming mail.

SNMP

(Simple Network Management Protocol) A widely-used network monitoring and control protocol. Data is passed from SNMP agents, which are hardware and/or software processes reporting activity in each network device (hub, router, bridge, etc.) to the workstation console used to oversee the network. The agents return information contained in a MIB (Management Information Base), which is a data structure that defines what is obtainable from the device and what can be controlled (turned off, on, etc.). Originating in the UNIX community, SNMP has become widely used on all major platforms.

TUDP

(User Datagram Protocol) A protocol within the TCP/IP protocol suite that is used in place of TCP when a reliable delivery is not required. For example, UDP is used for realtime audio and video traffic where lost packets are simply ignored, because there is no time to retransmit. If UDP is used and a reliable delivery is required, packet sequence checking and error notification must be written into the applications.

TCP/IP

Abbreviation for Transmission Control Protocol/Internet Protocol, the suite of communications protocols used to connect hosts on the Internet. TCP/IP uses several protocols, the two main ones being TCP and IP. TCP/IP is built into the UNIX operating system and is used by the Internet, making it the de facto standard for transmitting data over networks. Even network operating systems that have their own protocols, such as Netware, also support TCP/IP.

TCP provides transport functions, which ensures that the total amount of bytes sent is received correctly at the other end. UDP is an alternate transport that does not guarantee delivery. It is widely used for realtime voice and video transmissions where erroneous packets are not retransmitted. IP provides the routing mechanism. TCP/IP is a routable protocol, which means that all messages contain not only the address of the destination station, but the address of a destination network. This allows TCP/IP messages to be sent to multiple networks within an organisation or around the world, hence its use in the worldwide Internet. Every client and server in a TCP/IP network requires an IP address, which is either permanently assigned or dynamically assigned at startup.

Network Operating System

Network Operating System is also called a NOS. It is an operating system which includes software to communicate with other computers via a network and manages network resources. It manages multiple requests (inputs) concurrently and provides the security necessary in a multiuser environment. It may be a completely self-contained operating system, such as NetWare, UNIX and Windows NT, or it may require an existing operating system in order to function (LAN Server requires OS/2; LANtastic requires DOS, etc.).

One piece of the network operating system resides in each client machine and another resides in each server. It allows the remote drives on the server to be accessed as if they were local drives on the client machine. It allows the server to handle requests from the client to share files and applications as well as network devices such as printers, faxes and modems. In a peer-to-peer network, the network operating system allows each station to be both client and server. In a non-peer-to-peer network, dedicated servers are used, and files on a client machine cannot be retrieved by other users. In networks of PCs, NetWare is the most widely used network operating system. Windows NT, Windows for Workgroups, Windows 95/98, VINES, LAN Server, LAN Manager and LANtastic are also examples.

UNIX, combined with TCP/IP and NFS, VMS combined with DECnet, the Mac OS combined with AppleTalk, and SNA, combined with VTAM and NCP, also provide network operating system services.

Along with file and print services, a network operating system may also include directory services and a messaging system as well as network management and multiprotocol routing capabilities.

Examples of Operating Systems

NetWare

A family of network operating systems from Novell that support DOS, Windows, OS/2 and Macintosh clients. UNIX client support is available from third parties. NetWare is the largest installed base of LAN operating systems. Except for Personal NetWare and NetWare ELS, which were earlier peer-to-peer versions, NetWare is a stand-alone operating system that runs in the server. Until NetWare 5, which natively supports TCP/IP and Java, NetWare has always been a proprietary system. NetWare's native communications protocols are IPX, SPX and NCP. Its hard disks are formatted with the NetWare format, and although DOS and Windows applications reside in the server, they cannot be run in the server. All programs that run on a NetWare server are typically written in C and must be compiled using Novell libraries into executable files known as NetWare Loadable Modules (NLMs). Introduced in late 1998, the latest version is NetWare 5. Fully TCP/IP and Java based, Novell added significant enhancements to its flagship product, including a kernel that natively supports symmetric multiprocessing (SMP).

Introduced in 1993, NetWare 4 was the first NetWare version to use the much-acclaimed Novell Directory Services (NDS), which provides directory services for a global enterprise. Introduced in 1989 as NetWare 386 and then again in 1992 as NetWare 3.11, it was the first 32-bit version of NetWare, which has a limit of 250 concurrent users. Still being sold, it uses the Novell bindery which provides directory services for a single server unlike the global NDS directory. NetWare 2.x (originally Advanced NetWare 286 in 1985) ran in a 286 and supported up to 100 concurrent users.

UNIX

Pronounced yoo-niks, it is a multiuser, multitasking operating system that is widely used as the master control program in workstations and especially servers. Myriads of commercial applications run on UNIX servers, and most Web sites run under UNIX. There are many versions of UNIX, and, except for the PC world, where Windows dominates, almost every hardware vendor offers it either as its primary or secondary operating system. Sun has been singularly instrumental in commercialising UNIX with its Solaris OS (formerly SunOS). HP, SCO, IBM and Digital have also been major UNIX vendors and promoters.

UNIX is written in C. Both UNIX and C were developed by AT&T and freely distributed to government and academic institutions, causing it to be ported to a wider variety of machine families than any other operating system. As a result, UNIX became synonymous with "open systems."

UNIX is made up of the kernel, file system and shell (command line interface). The major shells are the Bourne shell (original), C shell and Korn shell. The UNIX vocabulary is exhaustive with more than 600 commands that manipulate data and text in every way conceivable. Many commands are cryptic, but just as Windows hid the DOS prompt, the Motif GUI presents a friendlier image to UNIX users.

Command UNIX DOS

— List directory ls dir

— Copy a file cp copy

— Delete a file rm del

— Rename a file mv rename

— Display contents cat type

— Print a file lpr print
— Check disk space df chkdsk
— Change directory cd cd

History of UNIX

UNIX was developed in 1969 by Ken Thompson at AT&T, who scaled down the sophisticated MULTICS operating system for the PDP-7. The named was coined for a single-user version (UNo) of "multIX." More work was done by Dennis Ritchie, and, by 1974, UNIX had matured into an efficient operating system primarily on PDP machines. UNIX became very popular in scientific and academic environments. Considerable enhancements were made to UNIX at the University of California at Berkeley, and versions of UNIX with the Berkeley extensions became widely used. By the late 1970s, commercial versions of UNIX, such as IS/1 and XENIX, became available.

In the early 1980s, AT&T began to consolidate the many UNIX versions into standards which evolved into System III and eventually System V. Before Divestiture, AT&T licensed UNIX to universities and other organisations, but was prohibited from outright marketing of the product. After divestiture, it began to market UNIX aggressively.

Advantages of UNIX

Even with its many versions, UNIX is widely used in mission critical applications for client/server and transaction processing systems. UNIX components are of world class standards. The TCP/IP transport protocol and SMTP e-mail protocol are de facto standards on the Internet. NFS allows files to be accessible across the network, NIS provides a "Yellow Pages" directory, Kerberos provides network security, and X Window lets users run applications on remote servers and view the results on their machines.

Windows NT

(Windows New Technology) An advanced 32-bit operating system from Microsoft for Intel x86 and Alpha CPUs. Support for the PowerPC and MIPS platforms was dropped. Introduced in 1993, NT does not use DOS, it is a self-contained operating system that runs 16-bit and 32-bit Windows applications as well as DOS applications.

There are actually two versions of Windows NT: Windows NT Server, designed to act as a server in networks, and Windows NT Workstation for stand-alone or client workstations. Features include peer-to-peer networking, preemptive multitasking, multithreading, multiprocessing, fault tolerance and support for the Unicode character set. NT provides extensive security features and continually tests the validity of application requests even after the application has been opened. Windows NT supports 2GB of virtual memory for applications and 2GB for its own use. Windows NT and Windows NT Workstation are the first and second releases of the client version. Windows NT Advanced Server (NTAS) and Windows NT Server (NTS) are first and second releases of the server version, which supports symmetric multiprocessing (SMP) and provides transaction processing for hundreds of online users. NT includes a dual boot feature.

NT Server is being widely implemented. NT's SMP capability takes advantage of Pentium Pro and Pentium II systems that contain two, four and more CPUs. As these multiprocessor systems become mainstream, NT competes squarely against RISC-based multiprocessor servers running UNIX. NT Workstation is also gaining market share in high-end desktop systems.

The last version of Windows NT with the Program Manager interface was Version 3.51. Introduced in the summer of 1996, Windows NT 4.0 contains the Windows 95 user interface. There are differences in some of the dialog boxes as NT contains features not available in Windows 95, and vice versa. NT 4.0 also includes Microsoft's DCOM interface that allows applications to be distributed across the network. NT 4.0 does not support Plug and Play, as does Windows 95. NT Server 4.0 comes with Microsoft's Internet Information Server (IIS), which provides Web server capability. NT Version 5.0, which was renamed Windows 2000, is expected in 1999. It adds Plug and Play support, Direct3D support, Active Directory, Zero Administration for Windows (ZAW) and other enhancements.

Version date/intro

— 3.1 July 1993

— 3.5 Sept 1994

— 3.51 Aug 1995

— 4.0 Aug 1996

— 2000 1999

Benefits of Network Operating System

A network operating system (NOS) causes a collection of independent computers to act as one system. A network operating system is analogous to a desktop operating system like DOS or OS/2, except it operates over more than one computer. Like DOS, a network operating system works behind the scenes to provide services for users and application programs. But instead of controlling the pieces of a single computer, a network operating system controls the operation of the network system, including who uses it, when they can use it, what they have access to, and which network resources are available.

At a basic level, the NOS allows LAN users to share files and peripherals such as disks and printers. Most NOSs do much more. They provide data integrity and security by keeping people out of certain resources and files. They have administrative tools to add, change, and remove users, computers, and peripherals from the network. They have troubleshooting tools to tell LAN managers what is happening on the network. They have internetworking support to tie multiple networks together.

Redirection

At the heart of the NOS is redirection. Redirection is taking something headed in one direction and making it go in a different direction. With redirection, an operating program does not know or care where its output is going. You are probably familiar with DOS redirection. For example, the DOS command DIR > FILENAME will redirect a directory listing to a file instead of to the screen. The ">" tells DOS to give the results of the command to the entity on the right. Network operating systems depend heavily on redirection, only in this case data is being redirected from one computer to another over the network cable, not over the PC's bus to local files or printers. Nevertheless, the operation is similar. If you type "COPY C: FILEA F:", FILEA will be copied from your local drive C: to the network drive F:. The NOS makes it appear to the COPY command that drive F: is local, when it really resides on another computer that is attached to the same network. The COPY command doesn't know or care that drive F: is across the network. It sends the file to DOS and the NOS reroutes the file across the LAN to drive F:. Redirection can be done with printers and other peripherals. Thus, LPT1: or COM1: can be a network printer instead of a local printer and the NOS redirects file to these devices. With a NOS, users don't need to know about redirection; they just type the drive designator or print from their word processors as always.

Server Software

The computer with drive F: must expect data, if the output from the user's PC can be redirected successfully. To do this, it must make its drive available to network users. This is part of the NOS's function at the server.

A NOS is made of a redirector and a server. Not all machines need to run the server software, because not all computers need to share their resources. But all LAN workstations must run redirector software because every client has to be able to put data onto the network.

With some NOSs, the computer running the server software cannot be used as a workstation. This is called a dedicated server. Novell's NetWare uses this kind of setup almost exclusively (although the low-end NetWare Lite can use nondedicated servers). With some other NOSs, all workstations on the network can also be servers. This a nondedicated server setup. This approach is used by Sitka and Artisoft, among others.

The two server approaches have advantages and disadvantages. Nondedicated servers allow for more flexibility, since users can make resources available on their computers as necessary. However, a nondedicated server approach requires that the users are willing to take some administrative responsibility for their computers and it necessitates that they be somewhat LANliterate. Backing up the shared data, setting up security, and setting up access rights become more complicated and often become the responsibility of the user, not the administrator. Another drawback is that non-dedicated servers often suffer some performance degradation when being used simultaneously as a workstation and as a server.

Dedicated servers have the opposite advantages and disadvantages. They are easier to administer since all data is in one place. They are faster because they don't have a local user to serve. On the other hand, it is harder to make resources available on an ad hoc basis, since setting up a server is more difficult and timeconsuming.

If a dedicated server fails, all users are forced to stop working because all resources are centralised. Your choice of dedicated or non-dedicated operation will depend on the work your network is doing.

File Service

A file server's primary task is to make files available to users, although it also makes other resources available, including printers and plotters. File service allows users to share the files on a server. The server PC can make its whole disk, certain directories, or certain files available. The file server's hard disk becomes an extension of each user's PC. The NOS can let the LAN administrator determine which users are allowed to use which files, for example, keeping the mail clerk out of the payroll file. Suppose a user wants to use a file residing on the file server's hard disk. Drive F: is set to correspond with the file server's hard disk. The actual process of setting up virtual drives has several names, including mapping, mounting, and publishing.

Now, suppose a user wants to run WordPerfect. At the F: prompt, the user types "WP" to load WordPerfect. WordPerfect is loaded from the server over the network, and into the user's PC's memory. Meanwhile, other people can use WordPerfect from the file server (assuming there is a license for multiple users). WordPerfect makes sure no other user can get the document file being used by "locking" the file. With many applications, file locking allows other users to read the document but not edit it.

File service is an extension of the local PC. Applications work just as they would on a local PC. Some programs, however, have been designed to take advantage of the network, rather than just run on one. For example, some databases allow two users to edit the same table but not the same record and each user can see the other's changes.

The NOS provides much more than just file service; it provides security, administration, printer sharing, backup, and fault tolerance.

Server Operation

The server software makes a single-user computer into a multiuser machine. Instead of just one user, a server has many users. But we must qualify what we mean by "many users." A NOS allows many users to share the server's peripherals, printers, disks, and plotters, but it does not allow multiple users to share its processor. For now let's see how the file server allows users to share its peripherals.

In many cases, the file server is running the PC's native operating system (such as DOS or the Macintosh OS) as well as the NOS. When users' requests come in, the NOS receives and interprets them, then hands them to the operating system for execution. So if a request comes in to open a file, DOS opens the file and gives it to the NOS, which gives it to the user. If many users make requests at the same time, the NOS queues them and hands them to DOS one at a time. High-performance NOSs, including Novell NetWare, Banyan VINES, and Microsoft LAN Manager, do not run DOS in the file server. DOS is replaced with a multitasking operating system, thereby gaining a performance advantage; however, they lose some compatibility and require dedicated file servers. In NetWare's case, it is a proprietary OS. VINES runs Unix; LAN Manager currently runs OS/2 but eventually will use Windows NT.

OS/2 NOS

File service is a tremendous improvement over single-user operation, it pales in significance to the enhancements that come with the new NOSs based on multitasking operating systems such as IBM/Microsoft OS/2.

The biggest advantage of a multitasking operating system is a server can offer a task. That is, the server can offer its processor to other users while it is serving requests for files and printers. A fast server can be used to do onerous chores like program compiling, calculations, and database sorting. It also means new types of programs can take advantage of the server processor.

Privacy and Network Security

Since IP packets pass through a number of networks before arriving at their destination, it is possible that a user's conversations (via e- mail, telnet, talk etc.) may be monitored, or tampered with, in transit. This is a serious breach of privacy, but as yet no institutional structure exists to trace or punish the eavesdropper/saboteur. Indeed in many cases a user may be unaware that her transmissions are being recorded or modified. One of the ways a user can increase the privacy of her transmissions and reduce their vulnerability, is by using encryption programs.

Various public key encryption programs exist that allow a user to encrypt data with a 'public key' and transmit it over the Network, assured that even if it is intercepted it would be close to impossible to decode the data unless one had the 'private key' that corresponds to the 'public key'. The severe restrictions placed on the distribution of such powerful encryption techniques by the US government has stunted their application in popular network utilities. The result of this policy is that unless the user makes a determined effort to encrypt her data it is by default transmitted in unencoded form. This situation is equivalent to posting a postcard as opposed to a letter enclosed in a sealed envelope because no envelopes are available in the market and one has to construct them by hand.

Encryption techniques are not currently widely used by general users. This means that more often than not the user who fails to encrypt data must accept the initial violation of her privacy and can only take precautions against further breaches. Unless encryption programs are incorporated into the most popular network applications it seems doubtful that users will make the effort to encrypt individual transmissions. The current situation also retards the growth of the commercial Network as financial data cannot be transmitted over the network in a secure form. In a case where a user's privacy has been violated, the disciplining of the offender cannot be accomplished by the normal methods (ostracization, loss of status). It is true that people who invade other's privacy are not respected (though they may be feared) on the Net, but such persons are generally not very concerned about their status among the wider community.

The Network is such that users tend to have an unique view of the personalities they converse with. The anonymity the Network provides can be counter-productive if it lets people deceive each other, or worse makes the prospect too enticing to resist. Not only does a Net address provide extremely limited information about it's owner, addresses can be easily fabricated and misused. When to this is added the screen afforded by anonymous remailers, it becomes apparent that identity as understood on the Net is not the concept we are familiar with. This is not as alarming as it sounds. We can, after all make a few assumptions about any user, we know that the user is human and their style of writing and the opinions they defend afford much more information about them. So one can know an user well without ever having met her In The Real World (ITRW).

Anonymity brings up other questions though. It is generally accepted that posts routed through anonymous remailers are not accorded the same of respect posts from a user's own account are. To an extent this may be because anonymous users are perceived to be more aggressive and confrontational. The argument that some anon users do abuse the cloak their anonymity provides them would be difficult to disprove. Yet it is also true that the Net as a whole is a community of distance, and people tend to be more visceral on networks than they would be in 'the real world'.

The Network does break down accepted codes of conduct while creating new ones. Thus though a person can be much more effusive on the Net than she would be in a traditional social setting, it is also easier to ignore completely users whose opinion one is not interested in. Since these properties are inherent in a Network that does not support video conferencing, it would be ingenuous to argue that anonymous remailers created such an environment, they only aggravate an pre- existing phenomenon.

Security on Local Computer Systems

System security poses questions that are very different from those we have already discussed. A network or computer when compromised can directly affect the work of an organization and thousands of users. Keeping this in mind unauthorized use of a computer system can only be regarded as theft and must be treated as such. Thus the system administrator is justified in transgressing certain boundaries to protect her system from attack.

The position Cheswick and Bellovin take must be the starting point for any discussion on Network security. Since an organization owns it's computational resources, it must have the last word on the appropriate use of such resources. The 'cracker' who breaks into a system without the permission of its owner, is then infringing and violating the organization's property rights. In the interests of liberty without aggression, we cannot defend the actions of the cracker.

After establishing the injustice of the actions of the cracker we must examine the means available to discipline such a malefactor. As with all criminals, it is by definition a difficult task to catch a cracker on the Net. This difficulty is further compounded by the nature of the Network and it's propensity to support anonymity. If a system has been subjected to successful attacks, often the only thing a system administrator can do is attempt to patch the holes in her system's security. Apprehending the cracker is a tricky business and one is never assured of success. Even if the cracker is identified (by the site the attacks originate form) it is difficult to verify whether or not the attacking site is itself a front that has been compromised.

In the event that the cracker is tracked down to his home site, the security administrator can attempt to contact the remote site's system administrator but there is no assurance that action will be taken against the cracker. The situation can be thought of as being analogous to that between independent states (that may even have different laws) without extradition treaties. The multiplicity of hosts and the global nature of computer networks implies that users are subject to vastly different laws and regulations, and in some cases (due to the tendency of the Network to metamorphose rapidly) none at all.. As such, the victim of an attack can only trust that the system administrator at the other end will take the appropriate action against the cracker.

Arrangements —when they exist at all— between system administrators, are generally informal and operate on assumptions of reciprocity. This should not come as a surprise. No host on the Network can exist independently. A system administrator who refuses to cooperate with those who have been affected by the actions of her users can expect to have her access to many services revoked. Sites that are notorious for cracker activity are generally barred from services at many other sites (that are aware of the nature of the activities at cracker sites and their disrespect for the rights of others). Since this inconveniences other (innocent) users, the system administrator at a cracker site would probably be pressured by her own users as well.

Thus in the absence of an institutional framework, there exists a method to punish the offender and prompt local authorities to consider the matter seriously. This mechanism does not of course, aggress upon the offender, it works much like ostracism and in this respect gives every individual in the (global) Net community the opportunity to take actions she deems to be appropriate. There is no aggression on the part of the victims, and attempts are made to resolve disputes outside of the 'tit for tat' mindset. In an environment where all transactions are voluntary, outlaws are punished by being denied services available to others.

An issue related to that of privacy and security is the concern about the reliability of information on the Net. This is where the stark contrasts between the Net and broadcast media are most glaring. The network confers on each and every user the ability to disseminate information and her views to an exceptionally wide audience at a very low cost. The Net has turned all of its users into publishers capable of making their work available to the entire Network community. This then brings up the question of who id to be responsible for dangerously incorrect information or even viruses and other malignant computer programs.

Traditional laws cannot safely be applied to this environment as on-line publications and writing are generally not subjected to the sort of scrutiny traditional broadcast media are. In a sense the ease with which information can be distributed and corrected, reduces the cost of dissemination and correction thereby lulling the provider into complacency. This may not be very significant on a Usenet newsgroup where knowledgeable persons are eager to correct the mistakes of others, but can pose a problem on the World Wide Web where one is unsure of the extent to which the erroneous data has traveled and where there is no continuing discourse on the topic.

A correction or revision of information or a program may reach the user too late. In cases where a publicly distributed computer program may be used as a "Trojan horse", actual damage can be done by the provider (knowingly or innocently). As it is, the user is expected to treat every transmission with caution and a healthy skepticism, much as one would in the real world. Yet a degree of gradation has evolved so that information from certain sites is deemed trustworthy and software from reputable institutions is understood to be safe. Since the Net unleashes the power to inform, information regarding changes in an institution's stance towards quality also travels very quickly.

It can be seen then, that the network is not a place for anyone to be sinecure. The responsibility to place sufficient guards on her own property rests with the individual user or organization. Even in the event of a security breach, the user or system administrator must often rely on her own good-will and capabilities in tracking down the cracker. This is the situation today. It may change in the future as computer networks are subjected to laws similar to those protecting one's privacy on telephone and postal services. We must however, when implementing these laws acknowledge that an internal system to deal with crackers and irresponsible users already exists, and as far as possible enact only those laws that compliment this order rather than attempt to replace it.

A related, though perhaps less exciting, problem is that of plagiarism and acknowledgment on Networks. It is conceivable that ideas and information may be plagiarized without the author's consent. The global nature of the Network certainly makes such a thing easier than ever to accomplish and extremely difficult to control. We can treat such incidents as plagiarism and the Net community, by applying the system of morals and ethics that has developed within traditional publications, does take such incidence seriously. There is now a growing awareness of the distinction between e-mail, mailing lists and Usenet newsgroups as far as the privacy they afford is concerned.

An E-mail message is generally considered confidential and though it is possible to involve users without the consent of all parties engaged in the correspondence (with blind cc for instance), this practice is considered impolite. A similar convention applies to mailing lists (especially moderated and restricted lists) which are by their nature considered closed discussions among users who trust one another. It is considered good netiquette to inform the participants of a mailing list before bringing non-subscribers into the discussion. A very different perception is held of Usenet.

Usenet newsgroups are freely accessible and as such are public discussions. Newsgroup contributions are considered to lie in the public domain as they are posted on a global bulletin board by their author. An exception may be made when a particular thread involves the administration of a particular newsgroup. In such a case to cross- post the thread to other newsgroups may change the structure of the forum and skew the discussion. There exist however, social conventions that deal with these issues and it may not be necessary to regulate or adjudicate such discussions. In most cases the issue is resolved by the participants themselves, without the necessity of an outside arbiter.

Views of Networks

Users and network administrators typically have different views of their networks. Users can share printers and some servers from a workgroup, which usually means they are in the same geographic location and are on the same LAN, whereas a Network Administrator is responsible to keep that network up and running. A community of interest has less of a connection of being in a local area, and should be thought of as a set of arbitrarily located users who share a set of servers, and possibly also communicate via peer-to-peer technologies.

Network administrators can see networks from both physical and logical perspectives. The physical perspective involves geographic locations, physical cabling, and the network elements (e.g., routers, bridges and application layer gateways) that interconnect the physical media. Logical networks, called, in the TCP/IP architecture, subnets, map onto one or more physical media. For example, a common practice in a campus of buildings is to make a set of LAN cables in each building appear to be a common subnet, using virtual LAN (VLAN) technology.

Both users and administrators will be aware, to varying extents, of the trust and scope characteristics of a network. Again using TCP/IP architectural terminology, an intranet is a community of interest under private administration usually by an enterprise, and is only accessible by authorized users (e.g. employees). Intranets do not have to be connected to the Internet, but generally have a limited connection. An extranet is an extension of an intranet that allows secure communications to users outside of the intranet (e.g. business partners, customers).

Unofficially, the Internet is the set of users, enterprises, and content providers that are interconnected by Internet Service Providers (ISP). From an engineering viewpoint, the Internet is the set of subnets, and aggregates of subnets, which share the registered IP address space and exchange information about the reachability of those IP addresses using the Border Gateway Protocol. Typically, the human-readable names of servers are translated to IP addresses, transparently to users, via the directory function of the Domain Name System (DNS).

Over the Internet, there can be business-to-business (B2B), business-to-consumer (B2C) and consumer-to-consumer (C2C) communications. Especially when money or sensitive information is exchanged, the communications are apt to be secured by some form of communications security mechanism. Intranets and extranets can be securely superimposed onto the Internet, without any access by general Internet users and administrators, using secure Virtual Private Network (VPN) technology.

References

Andrew S. Tanenbaum, 2006. *Computer Networks*, Fourth Edition, Pearson Education.

Kurose James F and Keith W. Ross. 2005. *Computer Networking: A Top-Down Approach Featuring the Internet*, Pearson Education .

Wendell Odom,Rus Healy, Denise Donohue. 2010. *CCIE Routing and Switching.* Indianapolis, IN: Cisco Press

William Stallings, 2004. *Computer Networking with Internet Protocols and Technology*, Pearson Education.

6

The Internet

The Internet is a global system of interconnected computer networks that use the standard Internet protocol suite (often called TCP/IP, although not all applications use TCP) to serve billions of users worldwide. It is a network of networks that consists of millions of private, public, academic, business, and government networks, of local to global scope, that are linked by a broad array of electronic, wireless and optical networking technologies. The Internet carries an extensive range of information resources and services, such as the inter-linked hypertext documents of the World Wide Web (WWW) and the infrastructure to support email.

Most traditional communications media including telephone, music, film, and television are being reshaped or redefined by the Internet, giving birth to new services such as Voice over Internet Protocol (VoIP) and Internet Protocol Television (IPTV). Newspaper, book and other print publishing are adapting to Web site technology, or are reshaped into blogging and web feeds. The Internet has enabled and accelerated new forms of human interactions through instant messaging, Internet forums, and social networking. Online shopping has boomed both for major retail outlets and small artisans and traders. Business-to-business and financial services on the Internet affect supply chains across entire industries.

The origins of the Internet reach back to research of the 1960s, commissioned by the United States government to build robust, fault-tolerant, and distributed computer networks. The funding of a new U.S. backbone by the National Science Foundation in the 1980s, as well as private funding for other commercial backbones, led to worldwide participation in the development of new networking technologies, and the merger of many networks. The commercialization of what was by the 1990s an international network resulted in its popularization and incorporation into virtually every aspect of modern human life. As of 2011 more than 2.2 billion people—nearly a third of Earth's Human population—used the services of the Internet.

The Internet has no centralized governance in either technological implementation or policies for access and usage; each constituent network sets its own standards. Only the overreaching definitions of the two principal name spaces in the Internet, the Internet Protocol address space and the Domain Name System, are directed by a maintainer organization, the Internet Corporation for Assigned Names and Numbers (ICANN). The technical underpinning and standardization of the core protocols (IPv4 and IPv6) is an activity of the Internet Engineering Task Force (IETF), a non-profit organization of loosely affiliated international participants that anyone may associate with by contributing technical expertise.

Internet Infrastructure

Better understanding of Internet infrastructures is an important element underpinning wider issues bearing on electronic commerce. They can provide a better understanding of the challenges for the private sector in upgrading infrastructure and of comparative national performance. They also provide an important input into a better understanding of how the Internet is becoming more critical for overall economic and social development in OECD countries.

Network Wizards define an Internet host as a domain name that has an associated IP address record. This would be any computer system connected to the Internet (via full or part-time, direct or dial-up connections), such as oecd.org and www.oecd.org. The Network Wizards survey includes all Top Level Domains (TLDs) and generic Top Level Domains (gTLDs) and is undertaken every six months. The RIPE survey is undertaken monthly but is limited to TLD registrations in their service area. While both surveys are much appreciated by the Internet community the results need to be qualified and have several limitations.

The first qualification that needs to be made is that host data do not indicate the total number of users who can access the Internet. The second caveat is that these surveys do not reach every host on the Internet, as access to some hosts is blocked by company fire-walls. Recognising the limitation of this second factor Network Wizards changed their methodology for the survey undertaken in January 1998 to enable access to a greater number of hosts. Notwithstanding this change, surveys of Internet hosts may only be interpreted as the minimum size of the 'public Internet', as it is impossible to determine the number of users accessing services via each host.

The Netcraft Web Server Survey is a survey of web server software usage on computers connected to the Internet. Netcraft collect and collate as many hostnames providing an http-service as their survey can find, and systematically poll each one with an HyperText Transfer Protocol (HTTP) request for the server name. A host name is the first part of a hosts' domain name. In the July 1998 survey Netcraft received responses from 2 594 622 web servers. The growth rate for the first half of 1998 was 41 per cent. Some 96 per cent of these servers are in the OECD area. By far the largest number of web servers are under.com which has 60 per cent of all web servers. As for Internet hosts, it is possible to provide penetration by domain and to weight this by the number of gTLD registrations.

The country providing the most responses, on a per capita basis, is Denmark. This is because there are a lot of small virtually hosted sites in Denmark. Netcraft report, that while this is a characteristic of many countries, it is particularly so in Denmark and the Netherlands. TeleDanmark and Cybercity operate two of the largest virtual hosting sites in Denmark. Internet surveys of hosts and servers provide one indicator of Internet development and may be used as one potential indicator of comparative Internet development between countries.

The main limitations are not reaching all hosts or servers, and the structure of the domain name system being such that there is no guarantee that all hosts under a particular domain are located in a certain geographic location. For example, the reachable hosts of a user in France, registering under a gTLD, would appear under domains such as.com or.net rather than.fr. Nor is it necessarily the fact that a host using a second level domain under.fr will be physically located in France. That being said the OECD's observations, from an series of traceroutes to Websites under TLDs, are that by far the majority of hosts using TLDs are located in the country concerned.

The availability of gTLD registrations by country presented the first possibility of redistributing Internet hosts under domain names such as.com to individual countries. If 5 per cent of the total gTLD registrations are from a particular country then 5 per cent of the total number of hosts surveyed under gTLDs are reallocated to that country. This methodology could, no doubt, be subject to a number of caveats.

Nevertheless it seems reasonable to assume that this approach gives a more accurate distribution of Internet hosts, in OECD countries, than allocating all hosts under gTLD registrations to the United States. The results of the weighted methodology are most striking in the case of Canada where, for mid 1997, there was a 72 per cent increase in the number of hosts over the number of hosts surveyed solely under.ca. Other countries recording significant increases at that time, albeit from smaller base numbers of hosts, were Turkey, Spain, Luxembourg and France.

All these countries recorded a relatively large increase in the number of hosts relative to the average OECD increase of 21 per cent. The countries for which this made very little difference are those where users mainly rely on national TLD registrations, such as Iceland, the Czech Republic, New Zealand, Poland and Finland. The Netcraft Server surveys also provide one of the best available indicators of the growth of electronic commerce on the Internet. Whereas the best known search engines only cover http sites, Netcraft also undertakes a secure socket layer (SSL) survey. The SSL protocol was developed by Netscape for encrypted transmission over TCP/IP networks.

It sets up a secure end-to-end link over which http or any other application protocol can operate. The most common application of SSL is https for ssl-encrypted http which enables electronic commerce to take place. In August 1998, Netcraft received responses from more than 424 000 web sites using encryption. However most of these responses are excluded, in terms of electronic commerce web sites, because they do not have third party certification. Sites without a third party certification are not expected to be engaging in electronic commerce because of the warning message that gets generated. The key element for electronic commerce is third party certification with matching certificate.

Netcraft say plausible reasons for the large number of responses, where the name in the certificate did not match the site's domain name, might include web sites run from virtual hosting configurations where the provider sets up all customers with https services, with customers buying certificates when they start to make use of the facilities. Netcraft adds that sites where the certificate issuer is not a known certificate authority, typically indicate that site has generated and signed its own certificate, which is acceptable for prototyping, or where trust is not required outside a limited group of people, such as a company, or collaborative project. This is likely to be more commonplace on internal networks than on externally visible Internet sites.

The major electronic commerce uses of secure server software are for encrypted credit card transactions over the Internet. The most common non-retail use of SSL is subscription access to privileged information. For example many of the leading United States investment banks disseminate research over SSL, and there are some applications for virtual private networks or closed access communities. By excluding sites without third party certification it is possible to get an indication of the number of electronic commerce sites in each OECD country.

Unlike the Internet host and general Netcraft web server survey, the SSL survey does not use the domain name system to categorise location but uses the actual address of the business. In August 1998, there were over 22,200 web sites engaged in electronic commerce via SSL in the OECD area. This number had grown by 128 per cent over the previous twelve months. The United States is a clear leader with three quarters of all electronic commerce sites but its overall share is falling as electronic commerce picks up speed in other countries.

Albeit starting from smaller bases, the number of electronic commerce sites grew by more than 300 per cent over the previous year in some OECD countries. Australia, Canada, and the United Kingdom have the highest number of electronic commerce Websites after the United States. In relative terms the United States also leads the OECD with more than six electronic commerce sites for every 100,000 inhabitants followed by Iceland, Australia and Canada.

Functions of Domain Name System (DNS)

The Domain Name System (DNS) maps Internet addresses. To function as part of the Internet a host needs a domain name that has an associated Internet Protocol (IP) address record. This includes any computer system connected to the Internet via full or part-time, direct or dial-up connections. DNS servers perform the necessary function of translating back and forth between names and numbers. These servers contain databases of IP addresses and corresponding domain names and they are interrogated each time a user wants to send an e-mail or request data over the World Wide Web. A top-level domain name (TLD) can either be an ISO country code or one of the generic top level domains.

To register a second level domain name or a third level domain name a user needs to apply to the domain name registry with the delegated authority for the TLD or gTLD. Some registries publish data on the number of registrations on a monthly basis while others publish intermittently. The Internet Hosts surveys undertaken by Network Wizards and RIPE also provide, as a by-product, an indicator of the number of registrations under each domain. The main importance of DNS indicators are that they can be used to inform discussions over the different policies and prices of TLD and gTLD registries.

Whether the registration process under a certain domain name is subject to industry self-regulation or government oversight, the availability of DNS data is important to ensure transparency in registration management for service providers, business users and consumers. This is particularly important in those cases where a monopoly or monopoly power exists in the registration of second and third level domain names. These tools can provide useful information for constructing certain Internet indicators. While this information is not always reliable, and some users would like to see additional information or functionality included, the various 'Whois?' databases are valuable resources for the Internet community.

The United States White Paper "A Proposal to Improve the Technical Management of Internet Domain Names and Addresses Discussion Draft", contained a number of suggestions for the type of information that should be included in domain registration databases and it will be up to the new DNS authority to work through guidelines in this area. Internet Protocol (IP) addresses are the numbers used to identify computers, or other devices, on a TCP/IP network. Networks using the TCP/IP protocol route messages based on the IP address of the destination.

The format of an IP address is a 32-bit numeric address written as four numbers separated by periods. Each number can be zero to 255. For example, 193.51.65.17 is one IP address used by the OECD. On a stand alone private TCP/IP network, IP addresses can be assigned at random as long as each one is unique to that network. If that private network connects to the Internet it requires a registered IP address to avoid duplication. The current version of IP (IP version 4 or IPv4), which was standardised in 1981, created a pool of 4 294 967 296 IPv4 addresses.

Originally blocks of IP addresses were directly allocated by the Internet Assigned Numbers Authority (IANA). While in some cases the IANA still makes direct assignments to organisations most allocations are now made to three regional bodies-APNIC (Asia-Pacific Network Information Center), ARIN (American Registry for Internet Numbers) and RIPE NCC (Reseaux IP Européens Network Co-ordination Centre). These organisations subsequently re-allocate or assign IP addresses to individual organisations, such as ISPs or national bodies co-ordinating IP address space for a certain country. As a general rule, end users receive IP assignments from their ISP. Notwithstanding this some ISPs receive addresses from upstream or backbone ISPs and some end users receive allocations directly from a registry or directly from the IANA.

The question of the whether the institutions allocating IP addresses at various levels have monopolies, in terms of guiding their self-governance practices, is important. Reform to the IANA is ongoing at the time of writing which takes into account its monopoly position. Whether the three regional registries have a total monopoly position is less clear because ISPs can, and do, apply to different registries (including allocations made to industry sectors such as the Cable Television Network based ISPs).

In addition, apart from convention, there may be nothing to stop an entity with a very large net block from reselling this resource in competition to the regional registries. Certainly many entities at the sub-regional registry level 'resell' IP numbers. The best sources of data on IP addresses are the three regional IP address registries. However, the policies and practices of each organisation mean the dissemination of IP allocation and assignment is carried out in different forms.

A further indicator to assist in the understanding of IP address allocations is work which has been undertaken by the Cooperative Association for Internet Data Analysis (CAIDA) to enable visualisation of IP address occupancy. Such depictions of the address space can also provide inputs for analysis of public policy (equity) issues, as well as information for evaluating engineering and operational aspects of the commercial Internet. Some of the main self-governance issues, relating to infrastructure, facing the Internet community are in the management of the existing IPv4 address space. These include:

— co-ordination of the three existing IP address registries;

— proposals to create new regional registries;

— the criteria for allocation of IP addresses;.

— the pricing structure used by registries for this resource.

Issues such as these can generate vigorous discussion within the Internet community and raise governance questions of a similar nature to the controversial DNS debate. Perhaps the main differences with IP address issues are that the main players are ISPs, rather than the much larger community which took an interest in DNS, and the absence of intellectual property concerns associated with trademarks and domain registration. Notwithstanding these differences, because the existing allocation of IP addresses are a monopoly, at least in one sense, the IANA and regional registries need to observe standards of openness, transparency and public accountability.

An essential part of this process, and a pre-requisite for self-governance, is the publication of data and indicators in a readily accessible form to the Internet community. While the management of IP addresses is best done by the private sector, policy makers need to have an understanding of the IP address system because of its potential to spill over into matters placed before governments. The most significant case in point is the ongoing problem of IP address portability and the competition questions raised. For example, if a relatively small ISP can not get a direct allocation of IP address space from a registry they need to borrow numbers from upstream ISPs.

If a small ISP then wishes to change its upstream (i.e. backbone provider), it has to return the borrowed IP numbers to the larger provider. In effect this means the smaller ISP has to renumber its network, which can be expensive and disruptive for its customers, if it wants to change backbone providers. This issue was raised in the context of the proposed merger of Worldcom and MCI which was considered by competition authorities in the United Sates and Europe. The concern voiced by some relatively small ISPs, arguing against this merger, was that if a dominant player in the backbone market abused that position they would face a high cost to shift providers because of the lack of number portability.

At the same time, the technical reasons for not encouraging portability for small address assignments and the attendant increase in the size of the routing table this would cause need to be borne in mind. According to some experts the portability problem will not be solved by the introduction of IPv6. Autonomous systems numbers (ASN, ASes or AS numbers) acts as a label for a set of IP addresses and are used by ISPs to specify the global routing policy for those IP addresses. On the Internet packets of data are passed between devices known as routers. Part of this process involves consultation of the routing table to determine the best onward path. A routing policy indicates 'reachability information', and hence enables traffic to pass between the networks of different ISPs.

For example, if an ISP has two connections to 'the Internet', it is often useful to spread traffic out over those two links. This policy is then propagated in order to insure that traffic (both incoming and outgoing) flows in the correct direction and that reachability of the addresses of an ISP providing services can be assured. In practice this means that most end users and small ISPs do not receive an individual AS number but instead use their upstream service provider's AS number. If an ISP only has one connection to the Internet (i.e. it is singly homed), it only has one way in which traffic can flow, thus it does not need an AS number.

However, the downstream ISP will still be exchanging traffic with the Internet, albeit under the AS number of their upstream provider. The distinction between being single-homed and multi-homed refers to the number of connections a small ISP, or content provider, has to the Internet via different backbone infrastructure providers. If an organisation is connected to the Internet via one ISP it is single-homed, but if it receives connections from two, or more, ISPs it is multi-homed. Internet backbone routers need to maintain the complete routing information for the Internet.

As the Internet has expanded, so too have the number of routes between AS numbers. This rapid growth has been a cause for concern among the Internet's technical community because it was felt that there was a limit to the amount of global routing information able to be accommodated by backbone routers. In mid-1997 an accident occurred which demonstrated the 'high-watermark' for the 'core Internet' and the fragility of the Internet. This was due to one Autonomous System accidentally re-announcing the entire Internet through itself resulting in a sudden doubling in the number of global routes to just over 80,000. As a result backbone routers around the world were overloaded as they did not have enough memory to cope with the additional routes. As these same routers tried to reboot, and establish peering, they once again received the information for more than 80 000 routes, and again ran out of memory.

This problem recurred until filters were manually installed or the originator of the excess routes fixed the source of the problem. This accident demonstrated both the current upper limit of the routing resource and the need for industry to have information and indicators to assist in self-regulating Internet traffic exchange. In other words, downstream ISPs, and smaller IP networks, used the AS number of their upstream backbone connection to the United States. However, the recent increase in allocation of AS numbers in the Asia Pacific has drawn some analysts to the conclusion that there is a corresponding increase in intra-national and intra-regional connectivity.

In the Asia Pacific this process is believed to have been spurred by the devaluation of a number of currencies (increasing the cost of international bandwidth) and the desire to improve network performance via localised (intra-national and intra-regional) traffic exchange. Further analysis of the policies for the AS numbers in appropriate databases, such those of APNIC and routing registries, could provide a tool for better understanding the amount of intra-national and intra-regional connectivity. Due to the nature of the Internet, measurement of performance across different networks can be much more difficult and controversial than the public switched telecommunication network (PSTN).

Measuring the performance within one network is relatively straightforward and many ISPs are comfortable enough to place near real-time indicators on the World Wide Web. This strikes an immediate contrast to the PSTN where such information was generally reserved for engineering purposes, and only made public in regulatory filings or annual reports. Accordingly, it is possible to see in near real time how, for example, MCI's Internet backbone is performing or the performance of links to Singapore Telecom's Internet exchange point (STIX).

Measurements include the latency of round trip times over different parts of networks and the loss of packets during times of congestion. Where the problematic aspect of measurement arises, is in constructing indicators to measure performance across different ISP networks, core Internet infrastructure (such as DNS) and the networks and sites of users. This can, of course, give rise to different interpretations of where the reasons for good and poor performance arise. There are several motivations for an increasing number of ISPs and Internet exchange points to put performance-related data on their Websites. Publication of such data can be used for marketing purposes and to provide customers with an additional service.

At the same time many of these indicators can be constructed by independent entities or by users themselves for their individual connections. As such there are a growing number of Websites that measure the "Internet Weather" from the perspective of their network or their Website. These are also companies that specialise in measuring the performance of particular backbone providers and ISPs, as well as services such as e-mail. In large part initiatives in the area of Internet network performance measures are best done by the private sector, albeit some government funded basic research institutions are currently playing an important role.

However if self-governance and industry co-operation evolves as the standard model for Internet traffic exchange, by contrast to the often heavily regulated interconnection in the PSTN, then the availability of performance indicators across different networks is essential. One issue that may arise for competition authorities or regulators is the claim that if one ISP held a majority share of the backbone market it might downgrade performance (i.e. traffic exchange) to other ISPs to encourage their customers to directly connect to the dominant player's network. Where there is a strongly competitive backbone market this problem should not arise.

However, in those countries where this is not the case, perhaps due to the existing or legacy effects of telecommunication monopolies, industry regulators may be called on to arbitrate very complex disputes between ISPs. On the other hand, the more indicators make transparent the relative performance of different networks the less likely such problems are to occur. Traceroutes enable users to follow a path taken by packets of data via intermediate routers between IP addresses.

A traceroute can be initiated from a user's PC, with a connection to the Internet, to a particular IP address (or a host/domain name such as www.oecd.org). The second option is to use one of the many sites on the World Wide Web which enable a user to traceroute from that web-site to an IP address using their browser. Boardwatch Magazine has compiled a comprehensive guide to traceroutes and how they work. Essentially traceroutes work by sending packets to an Internet address and getting intermediate routers to send a return message to the source.

The return message identifies the IP address of each router and the round trip time, in milliseconds, between the original router and each intermediate router. While the use of traceroutes as a tool for the Internet technical community to monitor the performance of peer networks is the leading application, traceroutes have also been used by some knowledgeable users to learn more about the performance of their ISP and their backbone provider. Users can learn, for example, how traffic to their nominated Websites traverses the Internet, and the network providers involved, in a much more transparent way than the public switched telecommunication network.

For policy makers traceroutes also represent an important tool for better understanding and informing debates on national and international traffic exchange. For example, some traceroutes demonstrate that traffic follows a certain path not because it may be the most efficient but because of the legacy or existence of infrastructure monopolies. In other words traffic may travel a certain route because capacity on a more logical route to provide connectivity is either unavailable or priced at uncompetitive rates. In addition, traceroutes to the most accessed Internet sites in different countries can be used as an indicator to better understand the relative positions of players in backbone markets.

Prior to consideration of this application, other interesting technical features, albeit sometimes erratic in terms of success, are:

— Traceroutes can indicate the capacity of a link between two routers either through the ISP incorporating this information (in the name of the router) or via traceroute software that endeavours to identify the capacity of each link.

— Traceroutes can be conducted between any two known hosts on some web sites.

— Some within the Internet community voluntarily incorporate latitude and longitude information in their routers and would like to see this practice generally adopted by all network managers.

— Users can employ traceroutes to provide an indication, or 'non-authoritative verification', of whether a Website is actually in the location specified on a Website or implied by a top level domain name or 'Whois? record'. In a number of cases bearing on electronic commerce or law enforcement and industry self-regulatory initiatives a user's first action is to use tools such as traceroutes to provide additional information about the source of data and possible recourse to ISPs or other authorities.

Worldcom and MCI forwarded the view that Internet revenues represented the best available indicator of market share. On the other hand critics of this view held that accurate data were not available either for the total Internet market or by relevant Internet market segment. Similarly, for policy makers considering the competition aspects of the merger, data were not available on Internet traffic such that an assessment of the market shares of various players could be constructed via that avenue. An alternative methodology is to use traceroutes to the most accessed Internet sites to gain a better understanding of backbone markets and Internet traffic exchange.

The basic idea is to undertake a series of traceroutes from a Website served by a certain backbone provider to the most sought after content and portals. It is then possible to see for how many of these sites a backbone provider can carry the traffic on an end-to-end basis and for how many of those sites they need to hand traffic to another provider. Together, a merger of Worldcom and MCI's backbone networks would have meant the new entity could carry the traffic internally to 62 of the leading 100 sites. Worthy of note is that this market position paralleled many of the market share estimates forwarded by those opposed to the merger of Internet backbones of these companies.

Yet, at the same time, the data also showed a high degree of multi-homing by major content providers in the United States with 53 of all sites being multi-homed and 35 of Worldcom customers being multi-homed. Another factor could be that the Websites in the United States bear far high traffic flows, than counterparts in other countries, and therefore require a higher degree of multi-homing. It might also be simply a sign that the Internet market in the United States is more mature than in other OECD countries.

One other potential indicator of the commercial maturity of the US Internet market relative to Australia, Germany, Sweden and the Netherlands is the high proportion of university Websites that figure in the top 100 Websites for these countries. In other words, users in universities still account for a high proportion of Internet traffic and accordingly lift the ranking of university Websites relative to commercial Websites. One implication of this might be to bear in mind that academic backbone networks, while continuing to grow and provide important services, will either commercialise to capture mass market growth or form an increasingly smaller part of the total backbone market.

There are also interesting features that apply to the four individual countries, albeit in each market a variation on the methodology has been explored. In Sweden a series of traceroutes was also conducted from CERN in Switzerland to test the market positions from outside the country. The results were that SwipNet, owned by Tele-2 provided carriage for 41 per cent of the commercial sites (i.e. excluding Sunet the University network). Telia and Telenordia provided carriage for a further 39 per cent of commercial sites. However, the most striking feature of this traceroute series was the amount of traffic between Denmark and Germany that traversed the United States.

For one quarter of the traceroutes from TeleDanmark, to the leading sites in Germany, traffic was initially exchanged via MCI's network in the United States and then found its way back to Germany via the Netherlands or other European countries. Whereas most large backbone providers which peer hand off traffic to another backbone network, as soon as the network determines they do not have a direct connection to the Website concerned and another path is available, Telstra's backbone network carries this traffic virtually the whole way.

Interestingly, this also leads to incoming international traffic being handed to Telstra, at an international exchange point, even if the ISP concerned has a direct backbone connection to Australia. Consider the series of traceroutes from CWIX (the Cable and Wireless US IP backbone network) which only carries traffic between CWIX and Optus (a Cable and Wireless Australian subsidiary), when it is to an Optus or Ozemail customer's Website. This international pattern of traffic is interesting in terms of Telstra's position on the financing of international infrastructure.

If CWIX (and Optus) carried the traffic from the United States to Australia, Optus would have had to pay Telstra for the domestic traffic exchange to reach the Telstra customer's Website. The interesting point is that it may be less expensive for CWIX to pass the traffic to MCI, than carry it directly to Australia and have it exchanged locally by Optus. There are a number of companies that seek to rank the most accessed sites on the World

Wide Web. Lists of the most accessed web sites are of interest to a variety of different groups.

For the companies wanting to advertise on the Internet it is useful to have an indication of the relative popularity of different Websites. For users such rankings are sometimes used as a directory service in different categories. For policy makers the main interest in Website rankings is the indication they might give for the location of the most popular content and as an input to determining the leading players providing access infrastructure to that content. The OECD has used Web21's lists of the most accessed Websites in different categories both to highlight to location of popular content, and its impact on traffic patterns, as well as the benefits being generated for pluralism and cultural diversity at a national and global level.

Such lists have also been used to determine which ISPs have the most significant backbone networks in terms of carriage of traffic to these high volume sites. A number of companies undertake studies endeavouring to rank Websites by the amount of traffic or visits they generate. Web21 produces the "Hot100" in different categories (e.g. news, sport, business, overall) and by different countries. Categories of content are useful when policy makers are endeavouring to better understand the location of popular content, such as webcasting radio stations and the potential impact on policy in respect to convergence between different communication platforms.

Country categories are useful for examination of traffic exchange, via traceroutes, between major backbone providers in national markets and their relationships with major international backbone markets. By combining tools such as traceroutes and lists of the most popular Websites indications can be constructed for different markets in terms of infrastructure availability, competition policy and the extent of trends such as multi-homing. Web21 collects data from a number of different sources but primarily use the logs from proxy servers and caches. These logs are sent on-line to Web21 on a daily basis and represent the surfing patterns of over 100 000 surfers world-wide.

Approximately 60 per cent of these users are in North America and 40 per cent outside the United States. Knowledge is another company which ranks web sites by getting users to download proprietary software via the Internet onto all computers they use to access the Web. These user's surfing patterns are then tracked by the company and made available on the net by different demographic categories. According to the Internet Advertising

Bureau (IAB) the top 10 Websites receive 67 per cent of total Internet advertising expenditures.

A portal is the term given to that part of a Website which acts as a gateway, or launch point, through which users navigate the World Wide Web. A portal can be the page a user sees each time they log onto the World Wide Web, such as the home page of an ISP or AOL (America Online). Accordingly, one measure of the usage of a portal is the number of subscribers served by a certain ISP, or other online service, although this is not precise as users can create their own portal or set their opening web page to another Website. These latter users will often define portals, such as Yahoo! or Excite, as their default Websites or use the default page of their browser (e.g. Netscape, Microsoft). These portals, which commenced service primary as search engines or directories have since moved to providing multiple applications designed to provide users with a variety of tools in one location (e.g. e-mail, news, personalised news, search engines, directories etc.).

Similarly, Netscape and Microsoft have incorporated an increasing number of portal-like functions into their browsers as well as selling hyper-text links and advertising space to other portals. Accordingly sites such as Yahoo and Excite are amongst the most accessed Websites on the Internet. It is this popularity, and the belief of many analysts that portals will become increasingly important to making the Internet more user friendly for electronic commerce, that has attracted increasing attention. Portals that were fairly recently in the preserve of sophisticated Internet users are now the best known sites on the Internet, a trend not missed by companies engaging in electronic commerce.

Another approach to investigating portals is to examine the number of hyper-text links to the gateway pages of these Websites and their origin. Many electronic commerce sites pay search engines, or other portals, a fee for each transaction referral. This can be done with, for example, 'click-through' banner advertising. In addition the total count of hyper-text links includes internal Website hyper-text links to a particular URL (e.g. back to main page hyper-text links, links to the web designer's page and so forth). It is also probable that if this measure became used as an indicator of popularity it would be relatively easy to distort.

Nicheworks has an interesting visualisation tool for Websites and links, showing the example of the Chicago Tribune's Website. By most measures

'Yahoo!' emerges as the most accessed Website on the Internet. Accordingly it is not a surprise that a count of the number of hyper-text links to http://www.yahoo.com/, reveals that it has the most links directed to it to on the World Wide Web. Excite and Lycos are the search engines that have the next highest number of hyper-text links followed by Alta Vista. Netscape and Microsoft also have Websites which attract some of highest numbers of links on the World Wide Web.

Apple and IBM make up the remainder of the group of companies with more than 100 000 hyper-text links to their main URLs. Results of other sites have been chosen to give an indication of the very large difference in the number of hyper-text links to 'new media' Websites and those to some traditional media. There is a vast difference in the number of hyper-text links to Websites needed to navigate the Internet (either via search and directory functions such as Yahoo! or equipment and software such as Netscape and Microsoft) for content and services than to the Websites of traditional print media.

As an aside it is also possible to use this tool to get an indication of the amount of 'adult content' relative to all content on the World Wide Web. Here it can be noted that the leading 100 Websites providing adult content make up only around 0.05 per cent of all content on the Internet. The proxy measure for content here is to count the number of outgoing URLs from these Websites relative to the total number of outgoing URLs from all domains. This does of course not deny the importance of using self regulatory tools, such as screening technologies, because of the ease with which adult material can be located intentionally or accidentally by children. However it is interesting to compare this figure with the amount of content under.edu (used by institutions of higher education in the United States) which makes up more than 16 per cent of all Internet content.

Taken together with the content under.us (used by schools in the United States) and the educational content under top level domains, the amount of educational content helps put a more balanced perspective on the different types of content available on the World Wide Web. It is also interesting to examine the number of hyper-text links to the traditional gatekeepers to the online world. Significantly only AT&T and NTT exceed 10 000 hyper-text links to their main URLs. In the not too distant past telecommunication carriers with government mandated monopolies owned every 'portal' from the actual line connection to the PSTN, through to every piece of customer equipment and the cover and contents of telephone directories.

Today, the majority of these markets have been liberalised in OECD countries, meaning that telecommunication carriers need to compete not only at the level of physical infrastructure but also for portals. Consider, for example, if as Internet telephony is increasingly incorporated into electronic commerce Websites, the additional communication revenue may accrue to the Internet telephony provider nominated by the owner of the portal. Even for the telecommunication carrier that becomes an ISP, and therefore has the opportunity to create a portal via the launch point for subscribers, there is no guarantee that these users will not opt for portals like Yahoo! and Netscape.

One indicator of the take up of different audio, video and other software tools on the World Wide Web is to use a search engine to count the number of applications. As Internet searches can be conducted under a particular top level domain name, by using HotBot, it is possible to get an indication of the use of these technologies. By including objects found under gTLDs and the TLDs associated with OECD Member countries, there appeared to be around 600,000 audio and more than 200,000 video applications in July 1998.

While the usual caveat applies, as to the allocations under gTLDs not being country specific, the data shown under TLDs might be taken to represent a very large sample for any given country. Accordingly by using search engines, analysts have one indication of the number of webcasting sites under a domain associated with a certain country. In future this may be an important source of information to assist policy makers dealing with convergence issues. This innovation created new pathways for electronic commerce and, by doing so, raised the question of whether indicators of the topography of linkages between domains could be useful for policy makers.

While the nature of the Internet's DNS makes it easier to apply some analytical tools to individual domains (i.e. yahoo.com), than to geographical areas (e.g..nl for the Netherlands), it is possible to construct matrices of the linkages between all the TLDs associated with OECD countries and gTLDs. The largest number of hyper-text links between top level domains are intra-domain links. However, an important factor is the relative development of the Internet in different countries. Attractive content has to exist under a certain domain, and be accessible (i.e. some users link to the same content under one domain rather than another based on infrastructure performance), before users will link to a URL. Linkages between domains provide a interesting indicator for further analysis of comparative performance even though allowances need to be made for gTLD registrations across different countries. The differences from traditional communication patterns may suggest more about Internet use and how electronic commerce is developing than the similarities.

Internet and Broadband Services

The development of the Internet is likely to be determined by the same growth of the myriad unpredictable commercial and social interactions that have fueled other communication services. The Internet has historically treated all packets equally, and pricing has been predominantly through flat monthly rates depending only on the size of access links, not on usage. Today, there is a strong momentum towards changing both of these principles. Although flat-rate continues to be the predominant form in which Internet access is sold, that form of pricing is unviable. Flat-rate pricing encourages waste and requires 20 percent of users who account for 80 percent of the traffic to be subsidised by other users and other forms of revenue. Furthermore, flat-rate pricing is incompatible with quality-differentiated services.

There is extensive work on development of multiple service levels, typically with some implied guarantee of performance, usually referred to as Quality of Service (QoS). If those are deployed, they will necessarily require pricing schemes more complicated than the standard flat rate ones. The driving force behind these developments is economic, based on the view that it is not feasible to provide satisfactory levels of quality for all transmissions at a reasonable cost. There is also concern (expressed in the quote above) about fairness of light users subsidising heavy ones under the current scheme.

While the momentum is towards introduction of QoS and more complicated pricing systems, there have always been voices arguing in favour of preserving the simplicity of both operations and pricing of the current Internet. Their counterarguments can be substantiated through evaluations of the costs of multiple service levels to end users, and on considerations of what utility users derive from data networks. Shortage of bandwidth is less of a problem than is often thought, and the general complexity of the system is a large barrier. An important observation behind many of the arguments in those papers is that what users really care about are fast transactions. As a result, whenever they can afford it, they run their links at low utilisations, since the main benefit they derive comes from the ability to burst at high speeds for short periods. In such an environment, flat rate pricing is most appropriate, and QoS does not help much. However unsound it may have seemed, unlimited local calling for a flat monthly fee for residential users has persisted in most of the United States for all of the 20th century. It seems worth exploring why it did, what effects that has had in comparison with other countries that adopted usage-sensitive pricing, why it is being reintroduced in many of those countries, and what conclusions we can draw from this about the likely future of the Internet.

History never repeats exactly, but there are frequent recurring themes that illuminate the present and hopefully can help make better choices for the future. While QoS and sophisticated pricing schemes are widely regarded as necessary for the future of the Internet, the trend in other communication services has almost uniformly been in the opposite direction. In long distance voice telephony, the most popular plans are the simple ones that are independent of time of day or distance. Even in the Internet arena, the trend so far has been largely towards simple flat rates.

A decade ago, the mass market online services, such as CompuServe, Prodigy, and AOL charged not just for minutes of connect time, but even for individual email messages. Email charges were eliminated first. By the middle 1990s, these services were also forced by subscriber complaints and competition from other ISPs to switch to unlimited access for a flat monthly fee. Although simplicity in pricing has been the general trend, it is important to recognise that there is persuasive logic to differentiated services. Even in the absence of quality differentiations, there are good reasons for pricing according to usage, as is done with electricity. Some of the arguments in this direction will be reviewed later. A major problem with such measures, though, is that usage sensitive charging, even with very low rates, lowers usage, often dramatically. Sometimes that is desirable. Western Europe is beginning to reintroduce forms of flat rate local phone calling precisely in order to stimulate Internet usage, which lags behind that in North America. Numerous examples are presented later to show that the best way to stimulate usage is to have flat rates.

Internet Services in Everyday Life

The Internet is expected to continue growing and to penetrate ever more deeply into our lives. Hence the competition among ISPs will be won by those who can entice their subscribers to make the fullest use of the services that are offered. Flat rates or approximations to them are among the most powerful tools in this competition. The general trend has been for each communication service to offer higher quality and lower prices. This has gone hand in hand with higher usage, and higher total revenues. A particularly telling instance of the trend towards simplification in communication is the tendency for prices to become distance-insensitive as a service evolves. The basic justification for the development of quality-differentiated service levels on the Internet is the perception that transmission and switching resources are expensive, and therefore high quality cannot be provided uniformly for all traffic. Yet that does not seem to have been a barrier for other communication services.

The distance dependence in pricing of mail, telegraph, telephone, and even data network services has historically decreased, or even disappeared completely. Long distance transmission is regarded as more valuable. Thus the decrease in distance dependence of pricing provides an even stronger argument than that of the hypothetical Martian. Perhaps an even more interesting point of view goes even deeper into an analysis of possible forms of pricing. A schedule of charges based simply on distance is already a major concession to the desire for simplicity, and away from optimality in either revenue extraction or in relating prices to costs.

A sounder procedure is to investigate more carefully what factors contributed to historical developments, and what influence they might have in the future. There are, after all, reasonable arguments against simplicity. Sophisticated strategies for quality differentiation and price discrimination are common. They are perhaps most noticeable in airline yield management, but are spreading to other areas. While there are sound arguments for quality differentiation and price discrimination, there are also economic arguments against them, based on economic models that favour bundling. Perhaps more important, such practices conflict with the strong preference that people have for simple pricing, especially for flat rate schemes.

There are many reasons for this preference. The insurance aspect is important. On the other hand, the search for optimal efficiency has usually been abandoned in favour of simplicity for small and frequent transactions. The long history illustrates how the decreasing prices of communication services, the fruit of improving technology and of economies of scale, have gone hand in hand with the trend towards greater simplicity. It seems reasonable to expect that a similar division between the quest for optimal efficiency on one hand, and simplicity on the other, will hold in the future. It will be easier for service providers to implement complicated pricing schemes. There will also be more tools helping consumers to navigate the maze of choices. However, both factors have been operating for a long time and that appears only to have strengthened people's desire for simplicity. The Internet is leading to an unprecedented proliferation of choices in goods and services, straining human capacity to deal with all the complexities. Therefore the balance between optimisation and simplicity is likely to remain where it has been, in spite of the developments in technology. That is a major reason that airline tickets are much more likely to be sold through complicated schemes than voice calls.

Even in the U.S., which is affluent, and has a large air transport industry, there is only about one round trip per person per year by air, at a cost of around $400. Thus airplane tickets are expensive and are bought relatively infrequently. Further, air transport is a mature industry, and costs are declining slowly. On the other hand, on average each U.S. inhabitant is engaged in about a dozen phone calls per day, each typically lasting about three minutes, and costing either nothing or a fraction of a dollar.

Human attention may be the ultimate scarce resource. The historical trend of increasing volumes and decreasing costs in communications is illustrated by comparing the purchase of a book today to the receipt of a letter in England in the the early 19th century. In 1834, about 5 letters were mailed per person per year in England. In most parts of the country, the recipient had to learn that a letter was waiting, go to the post office and pay a fee that averaged what about two hours of work cost. In contrast, in the U.S. of 1994, there were about two pieces of mail delivered per person per day. The cost of each piece was under two minutes' wages for an average worker.

Today, half a dozen years later, we are in the in the Amazon.com era. Those additional non-monetary book purchase costs can be decreased, as we can order books with a few keystrokes and mouse clicks, and have them delivered to our house a day or two later. Thus in terms of relative costs, books today are acquired more often and cost less than letters 170 years ago. Currently there are widespread expectations that " content," usually denoting professionally produced multimedia traffic, will dominate the Internet. However, even if the volume of multimedia traffic does exceed that of other transmissions, designing the network primarily for such traffic is likely to be a mistake.

Historical evidence, including current statistics, shows that people are willing to spend far more on point-to-point communication than on entertainment. Thus it is unlikely that professionally produced multimedia traffic will determine the evolution of the Internet. For example, in the early 19th century, U.S. postal policy was preoccupied with broadcast-type communication, namely through newspapers. Distribution of newspapers was subsidised to an extraordinary degree by regular letters, which brought in most of the money. Eventually public pressure forced policy makers to lower the prices of letters, which led to an upsurge of traffic.

Historical analogies are interesting, but may not always apply to a new technologies. Perhaps most important, prices have been increasing in recent years for most of the Internet. This is a seldom-noted phenomenon that is almost unique among modern high-tech industries in which demand has been increasing and technological progress has been stunningly fast. It is likely that this increase in transmission prices has been responsible for much of the demand for QoS and complicated pricing. Once prices start plummeting overall, as they have begun to do on some routes, it seems likely that providing more bandwidth will be seen to be more attractive than forcing complicated schemes on users. There are two other ways in which the Internet is often claimed to differ from other communication services.

One is in the possibility for a "tragedy of the commons," in which flat rate pricing leads users to saturate any links that are offered. The other is in the independence of traffic from distance. Service providers would act as insurance companies, using prior usage, demographics, and other information to fashion customised bundled offers that would meet the demands for simplicity and flexibility. Most standard works, such as, concentrate on optimising some conventional quantitative utility function, and treat customer desires as irrelevant, or else as irrational annoyances that interfere with clever and efficient schemes. It presents extensive historical and current data about customer preferences, explores the reasons for such preferences, and considers the effects of catering to them. It argues that however irrational such preferences may seem, they should be paid attention to. It shows that, generally, such a course of action is likely to be best for rapidly growing communication services like the Internet.

Technology is boosting the capacity of all networks. Therefore customers have to be encouraged to increase usage, to fill this growing capacity. Carriers that fail to persuade their customers to move on to higher bandwidth services will be left behind in the competitive race. Flat rate pricing is optimal in encouraging use. Standard works on telecommunications economics usually say nothing about the effect of flat rates on usage. This is obviously not the definitive history of communications. No work could possibly fill that role. In particular, practically no attention is devoted to technology, regulation, industrial organisation, and many other crucial aspects of communication services. Their effects are noted only through the decreasing prices and increasing quality and quantity that users experience. The main themes are growth of communication, the qualitative change this growth produces in how people treat communication, and the evolution of pricing.

The stress is on the user, and in particular on how quality and price differentiation have been used by service providers to influence consumer behaviour. For example, today telegraph company lawyers do not hire people to cut down the poles of a competitor. Also, railroads that compete for the same bottleneck mountain pass do not hire private armies that engage in shooting wars. A more substantial, but related, reason for concentrating on early stages of communication services is that they help clarify the evolution of those services. The contrasts between the early and mature stages of a technology bring into sharper contrasts the issues of how people treated those services, and what the role of pricing was.

The forces that have worked to produce simplicity in pricing in the past are likely to be at least as influential on the Internet. Optimal efficiency will be sacrificed in the interests of satisfying customers. Further, professionally produced "content" will not be the main factor determining the evolution of the Internet. Communications is likely to grow in revenues and importance, but will do so in often unpredictable ways, as society adopts the new tools to suit its needs. Price indices for the 19th century are rather imprecise, and in any case inflation-adjusted costs do not present an accurate picture of the burden of prices back when living standards were much lower.

Therefore some prices are stated in terms of hours of work required to pay that price.

Applications of Digital Convergence

The Internet is widely predicted to produce "digital convergence," in which computing, communications, and broadcasting all merge into a single stream of discrete bits carried on the same ubiquitous network. The popular images of convergence are heavily tinged with the flavour of Hollywood. "Content is king" is the universal buzzword, where content is usually taken to mean professionally prepared material such as books, movies, sports events, or music. The race is supposedly to determine which organisation or alliance will dominate in providing content to the users, ideally in advanced multimedia formats. A recent article concludes that "[the Internet] has become a mass medium used mostly by relatively passive consumers, and as such major content providers will dominate it".

The histories of the telegraph and the telephone show the same pattern. Usually many companies jumped in with unrealistic hopes, most failed, but the industry as a whole prospered. For example, the $187.5 billion of advertising industry revenues pays for almost all television broadcasting, and that provides much of the funding for the movie industry. Further, consumer expenditures on phone services are already contained completely in the general telephone industry figure.

Just the spending on phone services is higher than all advertising outlays. So say good-bye to all those plans for financing the Internet through advertising! Yes, advertising can help fund some services, but it will not provide the generous revenue streams that are needed to support a communications infrastructure as large as the phone system. To obtain the funding that many dot-coms seem to be planning on, it will be necessary to get contributions from more than advertising.

Ecommerce can help, but even that probably will not be enough, and it will be necessary to persuade people to pay for a large chunk of their communications. For example, sales of recorded music come to about $15 billion per year. If one eliminates the overhead costs of the physical distribution system for CDs, one could probably provide the artists with as much money as they make now, and the music labels with as much revenue for their central selection and promotional activities (and their profit) for under half of the $15 billion.

The postal system alone collects almost as much money as the entire movie industry, even though the latter benefits from large foreign sales. For all the publicity it attracts, entertainment is simply not all that large, because people are not willing to pay very much for it. The dream of the early 1990s of financing the "Information Superhighway" through "500 channels to the home on the cable TV network" was an obvious fantasy.

Content is not only a small part of the economy, it is often paid for indirectly. Well over two thirds of newspaper revenues, and almost all of broadcast TV and radio revenues come from advertising. Thus content is being given away in order to attract people to goods and services they are willing to pay for. Although spending for content has been rising, it has been doing so at a sedate pace, and is unlikely to explode. In particular, consumer spending on content is over 50% higher than on phone services. There is some issue of what one means by content, a question we will return to later. If we take a generous interpretation, we can come up with total content industry revenues comparable to the $256 billion that the telephone industry collected in 1997. However, comparing just the total revenues of those two industries is misleading.

In the case of the telephone industry, the $256 billion does include some service revenues as well as yellow pages advertising, but the overwhelming majority of that money is for simple transport of voice and data. The content industry as a whole, though, has to use its revenues to pay for content as well as the delivery of the content. The creative souls who command $20 million per movie know their value, and do manage to appropriate a lion's share of the profits from such enterprises, while keeping their risks lower than those of the investors. Similarly, the professionals who compile the Lexis database, or assemble and monitor Reuters' financial data feeds, have to be paid, just like movie actors and musicians.

Hence only some of the revenues of the content industry contribute to the communications infrastructure. That only a fraction of the revenues of the content industry go for delivery is an important point that has analogs in the context of ecommerce. Ecommerce is already big, and is exploding. A dollar of ecommerce transactions does not mean a dollar devoted to the network. Considerably more, but still only around $150 million, went for the servers, software development, and other information technology products and services that are needed to stay competitive in this rapidly changing field.

The contribution of ecommerce to communications is growing, but it has to be kept in perspective. Communications is far from being the largest segment of the economy. It is smaller than cars, housing, food, and especially medicine. It is also about two thirds the size of the primary and secondary education sectors, and comparable to the higher education enterprise. As a fraction of the U.S. economy, it has grown more than 15-fold over the last 150 years. The predominance of point-to-point communications spending is not new. That has been the historical pattern for ages. For example, in the early 19th century, almost all the revenues of the U.S. postal system came from letters. Yet about as many newspapers as letters were being delivered. The preoccupation of decision makers with content and broadcast communication is also not new.

Content of the Web

In the early 19th century, the explicit policy of the U.S. government was to promote wide dissemination of newspapers. They were regarded as the main tool for keeping citizenry informed and engaged in building a unified nation. The policy of the U.S. government to promote newspaper "content" at the expense of person-to-person communication through letters may or may not have been correct. However, there are reasonable arguments that the preoccupation with newspapers harmed the social and commercial development of the country by stifling circulation of the informal, non-content information that people cared about. In those days, the government understood clearly that what people were willing to pay for was letters, and that newspapers were being subsidised. However, this differential in willingness to pay does not seem to be understood as well today as it was then. Preoccupation with content has historically been common. For example, it was often thought that one of the principal uses of the telephone would be in broadcasting. In the end, though, the phone emerged as the prototypical example of point-to-point communication. Radio initially evolved in the other direction. It started out as a point-to-point communication technology, a wireless telegraph. After about two decades of experimentation, it became primarily a broadcast medium.

Further, in the last few decades, with the development of cellular services, radio transmission has started to move back to its roots as a point-to-point communications service. A skeptical reader might say that all this historical stuff is amusing but irrelevant. Studies of the Internet regularly find that Web traffic makes up 60 to 80% of the bytes that are transmitted. Certainly most of the commercial development effort on the Internet and almost all the attention are devoted to content. All argue that the dazzling success of the Web has created a misleading picture of what the Internet is, or is likely to evolve towards.

The content is not king of the Web. Most of the traffic on the Internet is corporate. It is likely that in early 2000, no more than a quarter of the volume went to residential users. Intranet traffic appears to be much less heavily biased towards the Web than that of private individuals. Furthermore, even the traffic that appears to be Web-based frequently represents a variety of database transactions that are not properly speaking "content." Because browsers are a user-friendly tool that is ubiquitous, a multitude of services have been squeezed into a Web framework, and help perpetuate the image of the Internet as primarily a content-delivery mechanism. Even if content were king on the Web now, the Web is not king of the Internet. This may again seem absurd, especially in view of the statistics quoted above, that most of the Internet traffic is Web transfers. Content certainly dominated in terms of volume of data.

Newspapers sent by mail weighed about 20 times as much as letters. Further, the density of printed matter is higher than of handwriting, and a typical copy of a newspaper was likely read many more times than a typical letter. Hence newspaper "content" was probably delivering at least a hundred times as much information as letters. On the Internet in 2000, it is email that is king, even if its volume is small. Today, Web traffic dominates the Internet in volume, with about 20 times as many bytes as email. Even a decade ago, before the Web, email typically accounted for under 10% of Internet volume. Yet email has been and continues to be the real "killer app" of the Internet.

The ARPANET (the progenitor of the Internet) was built primarily to connect computers. Yet email quickly emerged as the application that mattered the most to users, even in the early days of the network. This was much to the surprise of the system's designers. More recent surveys show that email is still the most valuable service. Ask people whether they would rather give up email or the phone, and the responses will typically be split. However, when a similar choice is offered between the Web and email, there is no contest. This is true for both individuals and large organisations. Intranets are all the rage, but it is email that makes enterprises run. The perception that content dominates the Internet is fueled by studies such as, which show the winner-take-all phenomenon, with a few sites dominating Web transactions among residential users. However, one should not read too much into such results. In the early 19th century postal system, studies of usage of information would undoubtedly have reached conclusions similar to those of. Most data from distant locations that people consumed came from newspapers. Further, circulation figures of individual newspapers probably followed the standard Zipf-type distribution, with the most popular papers attracting a disproportionately high fraction of readers.

To re-emphasise the importance of point-to-point communication in the online environment, consider the disappointing reception that WebTV has had. It seems that inexpensive Web browsing is not such a great attraction by itself. Their inadequate or even totally missing facilities for point-to-point communication appear to have been fatal errors. Next, consider the fates of CompuServe and Prodigy. Set up primarily for database access and online shopping, respectively, both were forced to emphasise communications and are now basically standard ISPs. The currently most successful of the public online services, AOL, started out as a game network. However, it simply cannot fund anything as large as AOL.) AOL survived and prospered because it was nimble. After several changes in strategy, it partially opened itself up to the Internet. While it has content of its own, and access to the Internet, the majority of the time its subscribers spend online is devoted to email and chat.

By focusing attention on centralised delivery of content, the Web may have prevented a proper appreciation of the importance of the often chaotic and generally unplannable point-to-point communications. The Web and the browsers may have played two main roles. One was to force online service providers to accept an open interoperable standard that made the entire Internet accessible for communications for everyone. The other was to introduce a user-friendly graphical interface for email, chat, and netnews, which made such communications easier. However, the Web is probably not as important to the Internet as is commonly thought. As an example of the relative value of content and simple pipes, note that the revenues of the entire cable TV industry in 1997 were only slightly higher than for the cell phone carriers. Yet cell phones currently provide primarily simple, low bandwidth pipes. By early 2000, even the gross revenues of radio telephony exceeded those of the cable TV industry. Further, a large chunk of cable revenues was devoted to paying for content, so in terms of basic network revenues, the cellular industry had pulled ahead even before 1997.

The comparison is even more favourable to low bandwidth wireless pipes when we go outside the U.S. In other industrialised countries, cell phones are often much more widely used, while cable TV penetration is almost always far lower than in the U.S. Thus on a worldwide scale, the comparison is skewed even more heavily against content. The cable TV industry does have excellent prospects for faster growth. However, that growth will surely come more from improved communication capabilities, and less from content.

The general conclusion is that content has been less important than point-to-point communication in the past, including the recent past involving the Internet. Historically, content has often dominated, and probably dominate now. There are arguments, to be presented later, that in the future, content will not provide most of the bits travelling on the Internet. That content is not king does not mean that content is unimportant in shaping political or social views. The attention paid to writers, and the political advertising on radio and TV, testify to the influence of content. This also is not a new phenomenon.

Content can be profitable. Numerous media companies are doing very well. Content can also be of value to a network, even aside from providing traffic for the network to carry. However, it is probably best to think of content as either catnip or icing on the cake; something to attract new users, or enhance user experience. That is what broadcast TV programs do for the advertisers who pay for them. That may also have been the main role of the Web and browsers in bringing more people to the Internet. Content is not king does say is that people are willing to pay far more for point-to-point communication than for the famed content. That is likely to be reflected in what kinds of networks are built, and which companies succeed. It inverts the usual ordering of priorities, making point-to-point communication central, and content secondary. The fights over control of movie distribution may be a distraction from the main business of communication. However, if the consumer places much higher value on personal communication than on content, the case for symmetrical connections becomes stronger. That may mean that fibber to the home may be justified sooner than expected. In the wireless arena, it suggests that for much of the next decade, the best strategy will be to emphasise regular voice transmission, supplemented by email and various low-bandwidth data transfers.

In general, content is distributed by "mass" or "broadcast" communications systems. Until a few decades ago, such services could be distinguished easily from "point-to-point" communications, which included first class letters and phone calls, and were specific to the people involved in the transaction. These two types of communications were sometimes combined during distribution, as in the postal system, which carried both letters and newspapers, in an early example of "convergence." However, there was a noticeable distinction in how these two types of communication were prepared, handled, perceived by the recipients, and in how much people were willing to pay for them.

Online Environment: Emerging Trends

During the last few decades, the distinction between point-to-point and broadcast communication began to blur. Computers allowed for the mass preparation of personalised letters offering credit cards, say. Answering machines and voice response systems led to machine-mediated point-to-point communications. Many of the standard phone calls access just such services, and the Internet is leading to increasing usage of them. These types of interactions will flourish on the Internet, and some will be merging with content, but they are more typical of the standard point-to-point communications.

One of the main lessons from the history of communications is that technologies are often adopted rapidly, but seldom at the astronomical rates that popular imagination associates with the Internet. This applies even to the Internet, where change tends to be less rapid than is often thought. Browsers were adopted rapidly. The first informal release of Mosaic took place in the spring of 1993, and in under two years, the majority of Internet traffic was Web-related. However, that was an exception. Other changes have been slower. Just consider Internet telephony, introduced back in 1995. It is finally beginning to have a noticeable effect, but it is far from dominant.

In general, the time a new technology takes to become widespread has not changed much in the last half century. Even technologies with compelling advantages tend to take a decade to dominate markets. Fax machines took about 10 years, from the introduction of the first inexpensive models until they became ubiquitous. Cell phones, one of the fastest growing industries, have taken about 15 years to reach their. Cable TV has taken three decades to reach about 60% of U.S. households. Music CDs and more recently DVDs show similar patterns, taking on the order of a decade to reach dominance. Aside from the unusually rapid ascendency of the browser-Web combination, fast change is usually associated with the presence of forcing agents. These can be either governments or a small number of key decision makers who can shift an industry's direction. An example of such forcing agents are the information technology managers at large banks and other enterprises.

A transition of voice phone traffic from circuit switched transmission to the Internet might occur rapidly for just this reason. The carriers might be able to implement it rapidly, since it could be done essentially invisibly to the end users, and the decisions would need to be made by only a few people. The decade-long diffusion periods listed above for consumer goods and services are due to the inertia of the millions of people who have to individually decide to adopt a new technology.

Most new products and services are in that category. Sociological changes are even slower, taking a generation or two. Normal change, with a simple shift in technology that offers serious advantages over older, more established competitor, takes on the order of a decade. The speed with which new technologies diffuse has a direct relevance for the question of whether content might be king of the Internet. It is possible to make a case that even if content is not king now, it might be king in the future, when convergence does occur. Some evidence for this can be derived from the comparison of cable TV and cell phone industries. Their revenues and growth rates were cited as demonstrating that point-to-point communication is more important than content.

If convergence moves delivery of most of the content to the Internet, and transport grabs two thirds of the total revenues from content, then the network will get a huge new source of revenue. Total content spending in the U.S. is comparable to that on the phone system, so two thirds of that would make a huge difference to network financing. This scenario is not totally implausible, since the current content distribution system is grossly inefficient.

The Internet offers a chance to reduce some of the inefficiencies of the current system. For the current dominant content producers, the real threat from the Internet is probably less from piracy, and more from disintermediation. New producers can come in and, unburdened by the high overheads and obsolete habits of established players, can offer better deals to both the creative talent and the consumers. The Internet is a disruptive technology, it does have unprecedented ability to emulate other delivery mechanisms, and we are already seeing rapid growth in music delivery on it. As an example of what can happen to even the most solid-seeming businesses, just consider the Encyclopaedia Britannica. In spite of having the greatest brand name and by far the best content in the encyclopedia field, it has been floundering, and has yet to find a viable business model. Thus dramatic changes are indeed possible in the electronic environment. Still, there are other, more substantial, arguments against the content thesis.

The Internet will surely have a major impact on the content industries. However, as was discussed above, consumers are slow to change their behavior. Even the Encyclopaedia Britannica has had a decade in which to flounder. Even in the business environment, adoption of new technologies that require a thorough re-engineering of all internal processes is slow. The business-to-consumer dot-coms have discovered this already, and the business-to-business ones are in the process of learning. Although we are living in the Internet era, fundamental change is not all that much faster than a century and a half ago. Slow adoption of new technologies means that convergence will be spread over a decade or more, and there will be continuing competition from traditional media, as well as increasing diversity of delivery mechanisms for content. This may mean that the writers and artists will get a bigger share of the pie. That appears to have been the trend over the last few decades, with movie actors and professional sport stars increasing their share of the revenues their work brings in.

There will certainly be a shift of revenue towards broadband services, but content distribution may not be the largest contributor to it. The main reason to question whether content will ever make giant contributions to network costs is that by the time convergence is likely to occur, at least a decade into the future, content transmission is likely to be a small fraction of total traffic. Further, most content will probably be distributed as ordinary file transfers, not in real-time mode. High prices could be charged for content distribution if content made up most of the traffic, or else if content required special transmission quality. Since neither of these conditions is likely to be satisfied, though, content will probably constitute just some of the huge number of large files, many encrypted, that will be flying around the network.

Internet backbone traffic appears to be about doubling each year. Further, advances in photonics and additional fiber deployment appear to allow for a doubling of network capacity each year for the next decade. In early 2000, Internet backbone traffic in the U.S. appeared to be less than a third of voice traffic in volume, but is likely to become larger by about 2002, with the transition point slightly later for the rest of the world. However, if voice traffic were to be packetised, it would almost surely be compressed, and then its volume would already be less than that of Internet traffic.

Broadcast TV would still overwhelm the Internet of the year 2000, but at 100% annual growth rates, it will not be too long before there is more than enough capacity to provide a high quality video channel for every person on Earth. This is because bandwidth that is likely to be required to satisfy any single person's real-time transmission needs will not be increasing fast. The versions of Moore's Law that hold in different industries operate at diverse rates. Microprocessors are doubling in computing power every 18 to 24 months, while fiber transmission capacity is doubling every year. On the other hand, display technology is advancing extremely slowly. Broadcast TV resolution has been static for several decades, and even the planned move to HDTV will require only a modest increase in bandwidth. Thus satisfying the needs for real-time multimedia transmission will not require much of the Internet's capacity.

Content is likely to form only a small fraction of Internet traffic for reasons explained. In addition, real-time delivery of content is likely to be an even smaller factor, for reasons discussed in much greater detail in. Transmission capacity is approximately doubling each year, which is a much faster rate of improvement than Moore's Law for semiconductors. However, hard disk storage capacity is also about doubling each year. Furthermore, that capacity is already huge. At the beginning of 2000, the U.S. Internet backbones appeared to be carrying about 12,000 TB (terabytes) of traffic each month. However, the total world hard disk capacity was 3,000,000 TB. Thus it would take about 20 years to transmit all the data on those disks over U.S. Internet backbones. The presence of huge local storage capacity in local PCs or cable TV setup boxes will make it much more attractive to send even content as files, not as real time transmissions.

There will be a growing volume for real time multimedia traffic, for applications such as videoconferencing. However, such applications are likely to be swamped by ordinary file transfers. The dominant mode of operation is likely to be fast download to local storage, fast transfer to whatever display device one wishes to use (often a mobile information appliance), and then playback. That is already the model we see emerging with MP3, Napster, and TiVo. The advantages of this model include the ability to implement it now, before the Internet can be made ready for real-time streaming media. It also accommodates gracefully the forecasted explosive growth in small mobile devices, which will often have small storage and low bandwidth over wireless links, and thus will be most useful if they can get data from local storage. This model also allows for easy integration with special hardware for intellectual property protection.

That real-time multimedia traffic would not dominate the Internet has been predicted several times in the past. It is an obvious conclusion from the rapid increase in traffic. The huge volume of local storage will surely stimulate the generation of non-content traffic. Both corporations and individuals so far have had no difficulty filling their disks with data. For firms, various databases will likely continue to proliferate and grow. For residential users, pictures are the leading candidates for filling those disks. Ease of use, lower cost, and instant gratification all stimulate use, and digital camera owners appear to be taking many more pictures than they ever did with regular film. The same electronic technology that is producing better disks and processors is also producing better cameras.

Historically, it appears that privately taken pictures have traditionally been the dominant source of data. Email and the Web may not be flexible enough. Yet the most successful communication technologies, the mail and the telephone, reached their full potential only when they embraced sociability and those "useless calls" as their goal. That seemingly idle chitchat not only provided direct revenues, but it encouraged the diffusion of the corresponding technology, and made it more useful for commercial and other applications. Such social interaction frequently function to grease the wheels of commerce. This work is largely the story of the development of the communications infrastructure from the point of view of the user. Usually an infrastructure is noted for being unnoticed; it is simply there, something we come to rely upon, do not have to think much about, and are horribly inconvenienced when it malfunctions.

Electricity, water, mail, and the phone are excellent examples. Yet they all took much effort to reach this stage. The key failing of the telegraph, as is detailed in later sections, is that it never became a true infrastructure component. It was a revolutionary technology, the "Victorian Internet," as one writer has called it, but it was too cumbersome and too expensive to attract much usage, and in particular never carried much of that "useless" social traffic that pushed the mail and the telephone to their eminent positions.

Traditionally, commercial users have subsidised residential ones. Sometimes this was done even involuntarily, as in higher rates dictated by carriers or by government regulators, and sometimes voluntarily, as in paying for toll-free 800 numbers. One of the purposes of this work is to examine the history of such subsidies. It appears probable that they will also play a large role on the Internet. The value of a broadcast network is usually regarded as proportional to the number of users in it. On the other hand, a point-to-point communication network is often said, by Metcalfe's Law, to have value proportional to the square of the number of member. This then leads to the conclusion that eventually, once a single network like the Internet reaches a large enough size, point-to-point communications will provide much higher value than broadcast.

With value proportional to the number of users, there would be few inherent advantages to an open network. The sum of the values of several completely or partially separate networks would be the same as of a unified network. On the other hand, if point-to-point communications were to dominate, and if Metcalfe's Law were to hold, there would be strong economic incentives to a unified network without barriers.

The Internet has done quite well without content, and can continue to flourish without it. Content will have a place on the Internet, possibly a substantial place. However, its place will likely be subordinate to that of business and personal communication. The explosive rise of the Internet is only the most recent chapter in a remarkable history of humanity becoming increasingly connected, the "annihilation of space and time," in a phrase going back at least to the early 19th century. However, growth rates rivaling that of the Internet have been observed occasionally before, for example in the development of the telegraph. The wireless industry is experiencing similar growth even now. A related one is to avoid the complexities of trying to measure the volume of information delivered by broadcast media. With point-to-point communication, it has often been true that the entire content was read, heard, or seen by the recipient. With broadcast communication, accounting is harder.

The communications industry as a whole has been expanding for centuries. Not only that, but most of the prominent services have continued growing. The tables in this section provide more eloquent testimony to this historical fact than words can. Even the venerable mail, often derisively called "snail mail," is still growing. Although there have been many predictions that mail would decline as a result of competition from the telegraph, the telephone, and more recently from email, it has continued to expand. Various new services have been added, such as money orders, home delivery, and parcel post. Even without the new services, the volume of letters alone has been increasing. Overall, total mail volume and spending have grown at impressive rates. Even in recent years prices have been decreasing, if we adjust them for increases in earnings. Moreover, this was not an isolated case of a frontier economy getting civilised.

In France, the postal system revenues are more than three times larger as a fraction of GDP than they were 150 years. In spite of this, mail is no longer the dominant communication technology, as it is dwarfed by the phone industry. Comparison of tables 3.1 and 3.4 shows the telephone passed the mail in revenues by 1920, and today is about four times as large. Adding the contributions of these two industries, we find that their role in the economy, measured as a fraction of GDP, has increased almost 20-fold over the last 150 years, from 0.2% to over 3.5%.The tide of increasing communications volumes has been powerful enough to lift most services. The telegraph did not stop the mail from growing. The telephone took a long time to eliminate the telegraph. Wireless telephony and the Internet have been growing explosively, yet wired telephony is still growing. Even fax use is increasing. Yet if anything should have been eliminated by the Internet, it surely should have been the fax.

Most messages sent by fax are typeset on a computer, printed out, and then fed into a fax machine. Still, the fax is ubiquitous, inexpensive, easy to use, and provides some features, such as signatures, that the Internet can and will eventually provide, but does not offer now. A technology that survived for longer than the fax is the electric telegraph. One can speculate that had the telephone not been invented, the telegraph might have spread more widely and become less expensive. Eventually it might have developed into something like our Internet, going from digital transmission directly to digital, bypassing the analog phone stage. There was intense development of new applications of telegraph technology in the second half of the 19th century for stock tickers as well as fire and home alarms. However, these developments came to nought.

The telephone started encroaching on the bread and butter business of the telegraph, and the latter could not compete effectively. The long survival of the telegraph suggests many lessons about rates of technological change. The Internet can emulate other communication technologies much better than the phone could emulate the telegraph, say. Still, the power of inertia should not be underestimated. The general progression for new technologies has been to start as an expensive and limited service. Then, as a result of technological improvements as well as economies of scale, prices usually declined and utilisation increased. Growth in usage was accomplished by development of new types of usage.

Some communication technologies, such as the electric telegraph and the telephone, went into public service right from the start. Initial usage was typically limited to special high-value commercial communications, since prices were high. In the Introduction, the sending of a letter in early 19th century England was compared (on the basis of earnings) to the purchase of a book today, costing about two hours of work. Telephone service initially was also extremely expensive. For example, a century ago, unlimited local service in New York City cost $20 per month. That is comparable to $2,000 per month currently, when one considers prices in relation to average earnings. The precise economic benefits of communication are a complicated subject beyond the scope of this work. However, they have consistently been high, often demonstrably so. The main reason telecommunications services are being liberalised and privatised around the world is to spur innovation and greater investments. Many careful studies have been conducted that documented the contribution of communications to speeding up economic growth. These benefits are easily appreciated by individuals.

A century ago, the highest penetration of phone service in the U.S. was in the Midwestern farming states. Even when there were no direct monetary savings, the savings in time from not making unneeded trips or arriving right on time were immense, and easy to understand for anyone. Once mail and phone service prices became low enough, social uses exploded. This enabled those two industries to increase their shares of the economy. The telegraph never moved beyond the high-priced and infrequent commercial use stage, and eventually died. The large and often immediately noticeable economic effects of communications meant that any changes in availability or pricing of services could have a large effect on the economy. Even regular mail played such a role. Much of the maneuvering in the U.S. Congress in the first half of the 19th century over postal rates had to do with the competitive positions of newspaper publishers in small towns versus those in large ones, and those in the South versus those in the Northeast. The interaction of mail with economic life developed gradually. This was largely because mail evolved gradually, and was a familiar service, simply performing better and less expensively what everyone could do, namely carry letters at the speed people could travel. Postal services did play a major role in the economy even in the early 19th century. However, mail was apparently not associated in the public eye with dramatic changes, at least not negative ones. The electric telegraph introduced a novel phenomenon.

The Internet is sometimes portrayed as a uniquely potent disruptive technology. Yet it is worth keeping in mind that the electric telegraph had a similar effect. In many ways, the telegraph was a greater innovation than the Internet. It was the first technology that effectively separated communication from transportation. It had a dramatic effect on science and technology, and it facilitated huge economic changes. The role of the telegraph in transforming the economy of the 19th century can be illustrated briefly by considering the history of the Rothschild banking dynasty. They were the world's largest financiers of railroads in the middle of the 19th century, and much of their wealth derived from that source. On the other hand, they were hostile to the telegraph, since it eroded their competitive advantage.

In the first half of the 19th century, the Rothschilds attained unprecedented dominance of international finance because they alone had a cohesive organisation covering all the major financial centres of Europe. The closely knit family could trust each other, and in addition they had a private communication system that provided them with news in advance of rivals. Most businesses in the mid-19th century had a choice of using the telegraph or dying, and using it effectively often meant large changes in modes of operation. This new technology was closely associated with new industries, especially the railroads, and new styles of management. It did lead to much more centralised control in both government and business affairs.

The establishment of a direct telegraph line between England and India in 1870 was an event of far-reaching importance. The delay in communication was a great advantage to the Government of India in so far as it of necessity left the initiation of policy in urgent matters to its own hands, and enabled it to confront the Secretary of State with accomplished facts. But all this was bound to change when the Secretary of State had to be kept constantly informed of the course of events in India, and was in a position to issue immediate orders. There is not doubt that the telegraph and the telephone had a great role in transforming society. However, many of the claims of dramatic changes caused by new communication technologies should be treated with caution. Corporate management was centralised, but there was an explosion in ranks of corporations, so some general measure of autonomy by managers may well have increased.

Many communication technologies led to predictions that they would lead to a new era of universal brotherhood and peace. For example, the International Telegraph Union, was the first permanent international technical organisation. There have been many predictions that the Internet would doom large corporations. In the absence of such costs, the factory worker, the garbage collector, the secretary, the software developer, and the salesman could just contract for their contributions to any given job in a decentralised market, without the need for a cumbersome bureaucracy. Since the Internet reduces transaction costs, it has often been predicted that it would eliminate the need for large organisations, and reduce the world to cottage enterprises.

The same phenomenon occurred in the second half of the 19th century, facilitated by the lower transaction and coordination costs offered by the telegraph. There does seem to be a strong tendency today to move away from vertical integration. That mode of operation appears to be replaced by enterprises concentrated in narrower horisontal layers of the economy, but operating on a worldwide scale. New technologies have often facilitated changes, but they were seldom the sole factors. Precise predictions have seldom been correct. Society has adopted the tools that were offered to it in a variety of often unexpected ways. The one factor that is undeniable, though, is that apparent from the tables in this section. Historically communication has been extremely highly valued. Individuals and enterprises have been devoting increasingly large fractions of their resources to it, and consuming rapidly rising volumes of it. The growth in volume of information has led to a change in how we react to it. Over the last two centuries we have moved from scarcity to surfeit. There have always been some complaints about too much information.

There are many worries that restrictive copyright laws will limit our access to data. However, in practice what we can see is an unprecedented abundance of information. With reduced marginal costs of distribution, content providers are fighting for eyeballs. Until a few years ago, you had to pay for a subscription to the New York Times, and also had to wait, often a long time, for your copy to be delivered. Today, you can get it for free on the Web soon after the editors decide on the final layout. When communication opportunities were rare, and transmission slow, it was natural that great care was taken with correspondence.

New communication technologies require new modes of acceptable usage. Today people complain about rudeness of cell phone users in restaurants or on airplanes, and wonder at the strange sights of cell phone users with headsets who seem to be talking to themselves. The phone also required development of new rules of etiquette. Even the wider use of mail for social communication led to “a burgeoning market for how-tomanuals to teach ordinary Americans the once-arcane custome of maintaining a correspondence with distant friends and family". For a communication service provider, the task is to make life for the customer as painless as possible. A seamless integration of the Internet into people’s lives should be the goal.

The Internet is the latest addition to the communication infrastructure that has been developing over the centuries. Most of the discussion in this work is limited to the 19th and 20th centuries. That was the period when the modern industrialised world was born, when the basic power, water, transportation, as well as communications infrastructures were built. Many similarities between the Internet and earlier communications services have been noted in earlier sections, and many more are presented in this one and later ones. However, there are also numerous similarities to other services, as well as common themes running through the combined history of all of them. For example, today’s carpal tunnel syndrome was known in the late 19th century as “glass arm” or “telegrapher’s cramp". As an example, most new technologies led to new forms of crime, and certainly to fears of crime. In late 1999 and early 2000, denial of service attacks on the Internet as well as other, real or potential problems, led to concerns that anonymity on the Internet might be harmful. The electric telegraph in France was initially limited to government use. Similarly, to prevent criminal elements from ‘using the telegraph to hatch their detestable plots with more ease and rapidity', the law demanded that private dispatches be sent un-coded and signed, and that the user declare his or her identity.

A common feature of all communication services has been that prices have usually had only a slight relation to costs. Average prices have, over long periods, decreased in line with the decline in costs. However, the structure of prices has traditionally included cross-subsidies. It is often difficult to tell when such subsidies exist. One could hope that at least the the economics of postal systems would be well understood. These systems are large, relatively slowly changing, and unusually open to public scrutiny.

The prominence of the Internet has tended to overshadow the rise of another great high-tech suc-cess, namely that of the wireless industry. There are more users of cell phones in the world than there are Internet users. Cell phones are a nice counterexample to the claim that is sometimes heard that metered services cannot grow fast. Wireless services have been growing rapidly, but have been metered from the beginning, and rates remain high. The U.S. cell phone industry collects on average about $0.25 per minute of wireless calls, as compared to $0.10 per minute for the wired long distance phone service. The widely varying costs of wired and radio transmission are reflected in the universal use of compression for voice calls over cell phones.

In recent years, as costs of transmission have decreased, use of compression even on very long underwater links has decreased. The slight disadvantages in complexity, delay, and so on that compression introduces have more than compensated for the substantial bandwidth savings that it offers. This is a nice example of how simplicity has triumphed over efficiency. On the other hand, compression on the radio link reflects the scarcity of bandwidth in that environment. Improvements in technology and greater spectrum allocations for communications will increase radio bandwidth. Still, this bandwidth is likely to remain orders of magnitude lower than that available on fiber optic backbones. Hence, even in the long run, although there may be plenty of bandwidth for voice calls, there will be a shortage for data, as mobile devices with powerful processors attempt to connect to big photonics pipes over relatively narrow radio channels.

Multimedia "content" will likely not be a major factor for many years. On the other hand, it seems that the prospects are excellent for stimulating much greater voice calling, since cell phones are used far less than wired ones. Digital messaging can be very useful in that regard. The scarcity of wireless bandwidth and its high price argue for quality and price differentiation. However, little of either is visible in the cellular industry. In the U.S., the cell phone owner pays for incoming calls. This imposes the risk of high charges not under the control of the user, and induces most people to keep their cell phones turned off except when they wish to make a call. The industry is trying to move to a "calling party pays" policy, but progress in that direction has been slow so far. Even if subscribers do not have to pay for incoming calls, other barriers will remain. Today, however, it might be better to raise the price of headsets, to move closer to a flat rate pricing system.

Internet Access

Common methods of Internet access in homes include dial-up, landline broadband (over coaxial cable, fiber optic or copper wires), Wi-Fi, satellite and 3G/4G technology cell phones. Public places to use the Internet include libraries and Internet cafes, where computers with Internet connections are available. There are also Internet access points in many public places such as airport halls and coffee shops, in some cases just for brief use while standing. Various terms are used, such as "public Internet kiosk", "public access terminal", and "Web payphone". Many hotels now also have public terminals, though these are usually fee-based. These terminals are widely accessed for various usage like ticket booking, bank deposit, online payment etc. Wi-Fi provides wireless access to computer networks, and therefore can do so to the Internet itself. Hotspots providing such access include Wi-Fi cafes, where would-be users need to bring their own wireless-enabled devices such as a laptop or PDA. These services may be free to all, free to customers only, or fee-based. A hotspot need not be limited to a confined location. A whole campus or park, or even an entire city can be enabled.

Grassroots efforts have led to wireless community networks. Commercial Wi-Fi services covering large city areas are in place in London, Vienna, Toronto, San Francisco, Philadelphia, Chicago and Pittsburgh. The Internet can then be accessed from such places as a park bench. Apart from Wi-Fi, there have been experiments with proprietary mobile wireless networks like Ricochet, various high-speed data services over cellular phone networks, and fixed wireless services. High-end mobile phones such as smartphones in general come with Internet access through the phone network. Web browsers such as Opera are available on these advanced handsets, which can also run a wide variety of other Internet software. More mobile phones have Internet access than PCs, though this is not as widely used. An Internet access provider and protocol matrix differentiates the methods used to get online.

An Internet blackout or outage can be caused by local signaling interruptions. Disruptions of submarine communications cables may cause blackouts or slowdowns to large areas, such as in the 2008 submarine cable disruption. Less-developed countries are more vulnerable due to a small number of high-capacity links. Land cables are also vulnerable, as in 2011 when a woman digging for scrap metal severed most connectivity for the nation of Armenia. Internet blackouts affecting almost entire countries can be achieved by governments as a form of Internet censorship, as in the blockage of the Internet in Egypt, whereby approximately 93% of networks were without access in 2011 in an attempt to stop mobilization for anti-government protests.

Social Impact

The Internet has enabled entirely new forms of social interaction, activities, and organizing, thanks to its basic features such as widespread usability and access. In the first decade of the 21st century, the first generation is raised with widespread availability of Internet connectivity, bringing consequences and concerns in areas such as personal privacy and identity, and distribution of copyrighted materials. These "digital natives" face a variety of challenges that were not present for prior generations.

Many people use the World Wide Web to access news, weather and sports reports, to plan and book vacations and to find out more about their interests. People use chat, messaging and email to make and stay in touch with friends worldwide, sometimes in the same way as some previously had pen pals. The Internet has seen a growing number of Web desktops, where users can access their files and settings via the Internet.

Social networking websites such as Facebook, Twitter, and MySpace have created new ways to socialize and interact. Users of these sites are able to add a wide variety of information to pages, to pursue common interests, and to connect with others. It is also possible to find existing acquaintances, to allow communication among existing groups of people. Sites like LinkedIn foster commercial and business connections. YouTube and Flickr specialize in users' videos and photographs.

The Internet has been a major outlet for leisure activity since its inception, with entertaining social experiments such as MUDs and MOOs being conducted on university servers, and humor-related Usenet groups receiving much traffic. Today, many Internet forums have sections devoted to games and funny videos; short cartoons in the form of Flash movies are also popular. Over 6 million people use blogs or message boards as a means of communication and for the sharing of ideas. The internet pornography and online gambling industries have taken advantage of the World Wide Web, and often provide a significant source of advertising revenue for other websites.

Another area of leisure activity on the Internet is multiplayer gaming. This form of recreation creates communities, where people of all ages and origins enjoy the fast-paced world of multiplayer games. These range from MMORPG to first-person shooters, from role-playing video games to online gambling. While online gaming has been around since the 1970s, modern modes of online gaming began with subscription services such as GameSpy and MPlayer. Non-subscribers were limited to certain types of game play or certain games. Many people use the Internet to access and download music, movies and other works for their enjoyment and relaxation.

Cybersectarianism is a new organizational form which involves: "highly dispersed small groups of practitioners that may remain largely anonymous within the larger social context and operate in relative secrecy, while still linked remotely to a larger network of believers who share a set of practices and texts, and often a common devotion to a particular leader. Overseas supporters provide funding and support; domestic practitioners distribute tracts, participate in acts of resistance, and share information on the internal situation with outsiders. Collectively, members and practitioners of such sects construct viable virtual communities of faith, exchanging personal testimonies and engaging in collective study via email, on-line chat rooms and web-based message boards.

References

Comer, Douglas. 2006. *The Internet book.* Prentice Hall. p. 64.

Hafner, Katie. 1998. *Where Wizards Stay Up Late: The Origins Of The Internet.* Simon & Schuster.

Manual Castells, 1996. *Rise of the Network Society*, Wiley-Blackwell.

Jesdanun, Anick. 2007. Internet Makeover? Some argue it's time". Seattletimes. nwsource.com. Retrieved 8 August 2011.

7

Database Management System

A Database Management System (DBMS) is a set of programs that enables you to store, modify, and extract information from a database, it also provides users with tools to add, delete, access, modify, and analyze data stored in one location. A group can access the data by using query and reporting tools that are part of the DBMS or by using application programs specifically written to access the data. DBMS's also provide the method for maintaining the integrity of stored data, running security and users access, and recovering information if the system fails. The information from a database can be presented in a variety of formats.

Most DBMSs include a report writer program that enables you to output data in the form of a report. Many DBMSs also include a graphics component that enables you to output information in the form of graphs and charts. Database and database management system are essential to all areas of business, they must be carefully managed. There are many different types of DBMSs, ranging from small systems that run on personal computers to huge systems that run on mainframes.

The following are examples of database applications: computerized library systems, flight reservation systems, and computerized parts inventory systems. It typically supports query languages, which are in fact high-level programming languages, dedicated database languages that considerably simplify writing database application programs. Database languages also simplify the database organization as well as retrieving and presenting information from it. A DBMS provides facilities for controlling data access, enforcing data integrity, managing concurrency control, and recovering the database after failures and restoring it from backup files, as well as maintaining database security.

Development of DBMS

Database servers are dedicated computers that hold the actual databases and run only the DBMS and related software. Database servers are usually multiprocessor computers, with generous memory and RAID disk arrays used for stable storage. Hardware database accelerators, connected to one or more servers via a high-speed channel, are also used in large volume transaction processing environments. DBMSs are found at the heart of most database applications. DBMSs may be built around a custom multitasking kernel with built-in networking support, but modern DBMSs typically rely on a standard operating system to provide these functions.

Databases have been in use since the earliest days of electronic computing. Unlike modern systems, which can be applied to widely different databases and needs, the vast majority of older systems were tightly linked to the custom databases in order to gain speed at the expense of flexibility. Originally DBMSs were found only in large organizations with the computer hardware needed to support large data sets.

1960s Navigational DBMS

As computers grew in speed and capability, a number of general-purpose database systems emerged; by the mid-1960s there were a number of such systems in commercial use. Interest in a standard began to grow, and Charles Bachman, author of one such product, the Integrated Data Store (IDS), founded the "Database Task Group" within CODASYL, the group responsible for the creation and standardization of COBOL. In 1971 they delivered their standard, which generally became known as the "Codasyl approach", and soon a number of commercial products based on this approach were made available.

The Codasyl approach was based on the "manual" navigation of a linked data set which was formed into a large network. When the database was first opened, the program was handed back a link to the first record in the database, which also contained pointers to other pieces of data. To find any particular record the programmer had to step through these pointers one at a time until the required record was returned. Simple queries like "find all the people in India" required the program to walk the entire data set and collect the matching results one by one. There was, essentially, no concept of "find" or "search". This may sound like a serious limitation today, but in an era when most data was stored on magnetic tape such operations were too expensive to contemplate anyway. Solutions were found to many of these problems. Prime Computer created a CODASYL compliant DBMS based entirely on B-Trees that circumvented the record by record problem by providing alternate access paths. They also added a query language that was very straightforward. Further, there is no reason that relational normalization concepts cannot be applied to CODASYL databases however, in the final tally, CODASYL was very complex and required significant training and effort to produce useful applications.

IBM also had their own DBMS system in 1968, known as IMS. IMS was a development of software written for the Apollo program on the System/360. IMS was generally similar in concept to Codasyl, but used a strict hierarchy for its model of data navigation instead of Codasyl's network model. Both concepts later became known as navigational databases due to the way data was accessed, and Bachman's 1973 Turing Award presentation was The Programmer as Navigator. IMS is classified as a hierarchical database. IDMS and CINCOM's TOTAL database are classified as network databases.

1970s Relational DBMS

Edgar Codd worked at IBM in San Jose, California, in one of their offshoot offices that was primarily involved in the development of hard disk systems. He was unhappy with the navigational model of the Codasyl approach, notably the lack of a "search" facility. In 1970, he wrote a number of papers that outlined a new approach to database construction that eventually culminated in the groundbreaking A Relational Model of Data for Large Shared Data Banks.

In this paper, he described a new system for storing and working with large databases. Instead of records being stored in some sort of linked list of free-form records as in Codasyl, Codd's idea was to use a "table" of fixed-length records. A linked-list system would be very inefficient when storing "sparse" databases where some of the data for any one record could be left empty. The relational model solved this by splitting the data into a series of normalized tables (or relations), with optional elements being moved out of the main table to where they would take up room only if needed.

For instance, a common use of a database system is to track information about users, their name, login information, various addresses and phone numbers. In the navigational approach all of these data would be placed in a single record, and unused items would simply not be placed in the database. In the relational approach, the data would be normalized into a user table, an address table and a phone number table (for instance). Records would be created in these optional tables only if the address or phone numbers were actually provided.

Linking the information back together is the key to this system. In the relational model, some bit of information was used as a "key", uniquely defining a particular record. When information was being collected about a user, information stored in the optional tables would be found by searching for this key. For instance, if the login name of a user is unique, addresses and phone numbers for that user would be recorded with the login name as its key. This simple "re-linking" of related data back into a single collection is something that traditional computer languages are not designed for.

Just as the navigational approach would require programs to loop in order to collect records, the relational approach would require loops to collect information about any one record. Codd's solution to the necessary looping was a set-oriented language, a suggestion that would later spawn the ubiquitous SQL. Using a branch of mathematics known as tuple calculus, he demonstrated that such a system could support all the operations of normal databases (inserting, updating etc.) as well as providing a simple system for finding and returning sets of data in a single operation.

Codd's paper was picked up by two people at Berkeley, Eugene Wong and Michael Stonebraker. They started a project known as INGRES using funding that had already been allocated for a geographical database project and student programmers to produce code. Beginning in 1973, INGRES delivered its first test products which were generally ready for widespread use in 1979. INGRES was similar to System R in a number of ways, including the use of a "language" for data access, known as QUEL. Over time, INGRES moved to the emerging SQL standard.

IBM itself did one test implementation of the relational model, PRTV, and a production one, Business System 12, both now discontinued. Honeywell wrote MRDS for Multics, and now there are two new implementations: Alphora Dataphor and Rel. Most other DBMS implementations usually called relational are actually SQL DBMSs.

In 1970, the University of Michigan began development of the MICRO Information Management System based on D.L. Childs' Set-Theoretic Data model. Micro was used to manage very large data sets by the US Department of Labor, the U.S. Environmental Protection Agency, and researchers from the University of Alberta, the University of Michigan, and Wayne State University. It ran on IBM mainframe computers using the Michigan Terminal System. The system remained in production until 1998.

Late-1970s SQL DBMS

IBM started working on a prototype system loosely based on Codd's concepts as System R in the early 1970s. The first version was ready in 1974/5, and work then started on multi-table systems in which the data could be split so that all of the data for a record (some of which is optional) did not have to be stored in a single large "chunk". Subsequent multi-user versions were tested by customers in 1978 and 1979, by which time a standardized query language – SQL – had been added. Codd's ideas were establishing themselves as both workable and superior to Codasyl, pushing IBM to develop a true production version of System R, known as SQL/DS, and, later, Database 2 (DB2).

Many of the people involved with INGRES became convinced of the future commercial success of such systems, and formed their own companies to commercialize the work but with an SQL interface. Sybase, Informix, NonStop SQL and eventually Ingres itself were all being sold as offshoots to the original INGRES product in the 1980s. Even Microsoft SQL Server is actually a re-built version of Sybase, and thus, INGRES. Only Larry Ellison's Oracle started from a different chain, based on IBM's papers on System R, and beat IBM to market when the first version was released in 1978.

Stonebraker went on to apply the lessons from INGRES to develop a new database, Postgres, which is now known as PostgreSQL. PostgreSQL is often used for global mission critical applications (the .org and .info domain name registries use it as their primary data store, as do many large companies and financial institutions).

In Sweden, Codd's paper was also read and Mimer SQL was developed from the mid-70s at Uppsala University. In 1984, this project was consolidated into an independent enterprise. In the early 1980s, Mimer introduced transaction handling for high robustness in applications, an idea that was subsequently implemented on most other DBMS.

1980s Object-oriented Databases

The 1980s, along with a rise in object oriented programming, saw a growth in how data in various databases were handled. Programmers and designers began to treat the data in their databases as objects. That is to say that if a person's data were in a database, that person's attributes, such as their address, phone number, and age, were now considered to belong to that person instead of being extraneous data. This allows for relations between data to be relations to objects and their attributes and not to individual fields.

Another big game changer for databases in the 1980s was the focus on increasing reliability and access speeds. In 1989, two professors from the University of Wisconsin at Madison published an article at an ACM associated conference outlining their methods on increasing database performance. The idea was to replicate specific important, and often queried information, and store it in a smaller temporary database that linked these key features back to the main database. This meant that a query could search the smaller database much quicker, rather than search the entire dataset. This eventually leads to the practice of indexing, which is used by almost every operating system from Windows to the system that operates Apple iPod devices.

21st Century NoSQL Databases

In the 21st century a new trend of NoSQL databases was started. Those non-relational databases are significantly different from the classic relational databases. They often do not require fixed table schemas, avoid join operations by storing denormalized data, and are designed to scale horizontally. Most of them can be classified as either key-value stores or document-oriented databases.

In recent years there was a high demand for massively distributed databases with high partition tolerance but according to the CAP theorem it is impossible for a distributed system to simultaneously provide consistency, availability and partition tolerance guarantees. A distributed system can satisfy any two of these guarantees at the same time, but not all three. For that reason many NoSQL databases are using what is called eventual consistency to provide both availability and partition tolerance guarantees with a maximum level of data consistency.

The most popular software in that category include: MongoDB, memcached, Redis, CouchDB, Apache Cassandra and HBase, that all are open-source software products.

XML Databases

A subset of NoSQL databases are XML databases. They all use industry standard XML data storage format. XML is open, machine-readable and cross-platform data format widely used for interoperability among different IT systems.

Software in this category include: BaseX, eXist, MarkLogic Server, MonetDB/XQuery, Sedna.

All XML databases can be attributed to document-oriented databases testing the additions.

Components of DBMS

DBMS engine accepts logical requests from various other DBMS subsystems, converts them into physical equivalents, and actually accesses the database and data dictionary as they exist on a storage device.

Data definition subsystem helps the user create and maintain the data dictionary and define the structure of the file in a database.

Data manipulation subsystem helps the user to add, change, and delete information in a database and query it for valuable information. Software tools within the data manipulation subsystem are most often the primary interface between user and the information contained in a database. It allows the user to specify its logical information requirements.

Application generation subsystem contains facilities to help users develop transaction-intensive applications. It usually requires that the user perform a detailed series of tasks to process a transaction. It facilitates easy-to-use data entry screens, programming languages, and interfaces.

Data administration subsystem helps users manage the overall database environment by providing facilities for backup and recovery, security management, query optimization, concurrency control, and change management.

Modeling Language

A modeling language is a data modeling language to define the schema of each database hosted in the DBMS, according to the DBMS database model. Database management systems (DBMS) are designed to use one of five database structures to provide simplistic access to information stored in databases. The five database structures are:

- the hierarchical model,
- the network model,
- the relational model,
- the multidimensional model
- the object relational model,
- the object oriented model,and
- the object model.

Inverted lists and other methods are also used. A given database management system may provide one or more of the five models. The optimal structure depends on the natural organization of the application's data, and on the application's requirements, which include transaction rate (speed), reliability, maintainability, scalability, and cost.

The Hierarchical Structure

The hierarchical structure was used in early mainframe DBMS. Records' relationships form a treelike model. This structure is simple but nonflexible because the relationship is confined to a one-to-many relationship. IBM's IMS system and the RDM Mobile are examples of a hierarchical database system with multiple hierarchies over the same data. RDM Mobile is a newly designed embedded database for a mobile computer system. The hierarchical structure is used primarily today for storing geographic information and file systems.

A hierarchical database model is a data model in which the data is organized into a tree-like structure. The structure allows representing information using parent/child relationships: each parent can have many children, but each child has only one parent (also known as a 1-to-many relationship). All attributes of a specific record are listed under an entity type.

In a database an entity type is the equivalent of a table. Each individual record is represented as a row, and each attribute as a column. Entity types are related to each other using 1:N mappings, also known as one-to-many relationships. This model is recognized as the first database model created by IBM in the 1960s.

Currently the most widely used hierarchical databases are IMS developed by IBM and Windows Registry by Microsoft.

The Network Structure

The network structure consists of more complex relationships. Unlike the hierarchical structure, it can relate to many records and accesses them by following one of several paths. In other words, this structure allows for many-to-many relationships.

For computer network models, see network topology, packet generation model and channel model.

The network model is a database model conceived as a flexible way of representing objects and their relationships. Its distinguishing feature is that the schema, viewed as a graph in which object types are nodes and relationship types are arcs, is not restricted to being a hierarchy or lattice.

The network model's original inventor was Charles Bachman, and it was developed into a standard specification published in 1969 by the CODASYL Consortium.

The Relational Structure

The relational structure is the most commonly used today. It is used by mainframe, midrange and microcomputer systems. It uses two-dimensional rows and columns to store data. The tables of records can be connected by common key values. While working for IBM, E.F. Codd designed this structure in 1972.The model is not easy for the end user to run queries with because it may require a complex combination of many tables

The Multidimensional Structure

The multidimensional structure is similar to the relational model. The dimensions of the cube-like model have data relating to elements in each cell. This structure gives a spreadsheet-like view of data. This structure is easy to maintain because records are stored as fundamental attributes—in the same way they are viewed—and the structure is easy to understand. Its high performance has made it the most popular database structure when it comes to enabling online analytical processing (OLAP).

The Object-oriented Structure

The object-oriented structure has the ability to handle graphics, pictures, voice and text, types of data, without difficultly unlike the other database structures. This structure is popular for multimedia Web-based applications. It was designed to work with object-oriented programming languages such as Java.

The dominant model in use today is the ad hoc one embedded in SQL, despite the objections of purists who believe this model is a corruption of the relational model since it violates several fundamental principles for the sake of practicality and performance. Many DBMSs also support the Open Database Connectivity API that supports a standard way for programmers to access the DBMS.

Before the database management approach, organizations relied on file processing systems to organize, store, and process data files. End users criticized file processing because the data is stored in many different files and each organized in a different way. Each file was specialized to be used with a specific application. File processing was bulky, costly and inflexible when it came to supplying needed data accurately and promptly. Data redundancy is an issue with the file processing system because the independent data files produce duplicate data so when updates were needed each separate file would need to be updated. Another issue is the lack of data integration. The data is dependent on other data to organize and store it. Lastly, there was not any consistency or standardization of the data in a file processing system which makes maintenance difficult. For these reasons, the database management approach was produced.

DATA STRUCTURE

Data structures (fields, records, files and objects) optimized to deal with very large amounts of data stored on a permanent data storage device (which implies relatively slow access compared to volatile main memory).

Database Query Language

A database query language and report object allows users to interactively interrogate the database, analyze its data and update it according to the users privileges on data. It also controls the security of the database. Data security prevents unauthorized users from viewing or updating the database. Using passwords, users are allowed access to the entire database or subsets of it called subschemas. For example, an employee database can contain all the data about an individual employee, but one group of users may be authorized to view only payroll data, while others are allowed access to only work history and medical data.

If the DBMS provides a way to interactively enter and update the database, as well as interrogate it, this capability allows for managing personal databases. However, it may not leave an audit trail of actions or provide the kinds of controls necessary in a multi-user organization. These controls are only available when a set of application programs are customized for each data entry and updating function. Data structure is a logically representation of relationships between individual elements and data.

Transaction Mechanism

A database transaction mechanism ideally guarantees ACID properties in order to ensure data integrity despite concurrent user accesses (concurrency control), and faults (fault tolerance). It also maintains the integrity of the data in the database. The DBMS can maintain the integrity of the database by not allowing more than one user to update the same record at the same time. The DBMS can help prevent duplicate records via unique index constraints; for example, no two customers with the same customer numbers (key fields) can be entered into the database.

Implementation of DBMS

A database management system (DBMS) is a system that allows to build and maintain databases, as well as to utilize their data and retrieve information from it. A DBMS defines the database type that it supports, as well as its functionality and operational capabilities. A DBMS provides the internal processes for external applications built on them. The end-users of some such specific application are usually exposed only to that application and do not directly interact with the DBMS. Thus end-users enjoy the effects of the underlying DBMS, but its internals are completely invisible to end-users. Database designers and database administrators interact with the DBMS through dedicated interfaces to build and maintain the applications' databases, and thus need some more knowledge and understanding about how DBMSs operate and the DBMSs' external interfaces and tuning parameters.

A DBMS consists of software that operates databases, providing storage, access, security, backup and other facilities to meet needed requirements. DBMSs can be categorized according to the database model(s) that they support, such as relational or XML, the type(s) of computer they support, such as a server cluster or a mobile phone, the query language(s) that access the database, such as SQL or XQuery, performance trade-offs, such as maximum scale or maximum speed or others. Some DBMSs cover more than one entry in these categories, e.g., supporting multiple query languages. Database software typically support the Open Database Connectivity (ODBC) standard which allows the database to integrate (to some extent) with other databases.

The development of a mature general-purpose DBMS typically takes several years and many man-years. Developers of DBMS typically update their product to follow and take advantage of progress in computer and storage technologies. Several DBMS products have been in on-going development since the 1970s-1980s. Since DBMSs comprise a significant economical market, computer and storage vendors often take into account DBMS requirements in their own development plans.

DBMS architecture specifies its components (including descriptions of their functions) and their interfaces. DBMS architecture is distinct from database architecture. The following are major DBMS components:

- *DBMS external interfaces* - They are the means to communicate with the DBMS (both ways, to and from the DBMS) to perform all the operations needed for the DBMS. These can be operations on a database, or operations to operate and manage the DBMS. For example:
    - Direct database operations: defining data types, assigning security levels, updating data, querying the database, etc.
    - Operations related to DBMS operation and management: backup and restore, database recovery, security monitoring, database storage allocation and database layout configuration monitoring, performance monitoring and tuning, etc.

  An external interface can be either a user interface (e.g., typically for a database administrator), or an application programming interface (API) used for communication between an application program and the DBMS.

- *Database language engines (or processors)* - Most operations upon databases are performed through expression in Database languages. Languages exist for data definition, data manipulation and queries (e.g., SQL), as well as for specifying various aspects of security, and more. Language expressions are fed into a DBMS through proper interfaces. A language engine processes the language expressions (by a compiler or language interpreter) to extract the intended database operations from the expression in a way that they can be executed by the DBMS.
- *Query optimizer* - Performs query optimization on every query to choose for it the most efficient query plan (a partial order (tree) of operations) to be executed to compute the query result.
- *Database engine* - Performs the received database operations on the database objects, typically at their higher-level representation.
- *Storage engine* - translates the operations to low-level operations on the storage bits. In some references the Storage engine is viewed as part of the Database engine.
- *Transaction engine* - for correctness and reliability purposes most DBMS internal operations are performed encapsulated in transactions. Transactions can also be specified externally to the DBMS to encapsulate a group of operations. The transaction engine tracks all the transactions and manages their execution according to the transaction rules (e.g., proper concurrency control, and proper commit or abort for each).
- DBMS management and operation component - Comprises many components that deal with all the DBMS management and operational aspects like performance monitoring and tuning, backup and restore, recovery from failure, security management and monitoring, database storage allocation and database storage layout monitoring, etc.

Database Storage

Database storage is the container of the physical materialization of a database. It comprises the Internal (physical) level in the database architecture. It also contains all the information needed (e.g., metadata, "data about the data", and internal data structures) to reconstruct the Conceptual level and External level from the Internal level when needed. It is not part of the DBMS but rather manipulated by the DBMS to manage the database that resides in it. Though typically accessed by a DBMS through the underlying Operating system (and often utilizing the operating systems' File systems as intermediates for storage layout), storage properties and configuration setting are extremely important for the efficient operation of the DBMS, and thus are closely maintained by database administrators. A DBMS, while in operation, always has its database residing in several types of storage (e.g., memory and external storage). The database data and the additional needed information, possibly in very large amounts, are coded into bits. Data typically reside in the storage in structures that look completely different from the way the data look in the conceptual and external levels, but in ways that attempt to optimize (the best possible) these levels' reconstruction when needed by users and programs, as well as for computing additional types of needed information from the data (e.g., when querying the database).

In principle the database storage can be viewed as a linear address space, where every bit of data has its unique address in this address space. Practically only a very small percentage of addresses is kept as initial reference points (which also requires storage), and most of the database data is accessed by indirection using displacement calculations (distance in bits from the reference points) and data structures which define access paths (using pointers) to all needed data in effective manner, optimized for the needed data access operations.

Data

Data is encoded by assigning a bit pattern to each language alphabet character, digit, other numerical patterns, and multimedia object. Many standards exist for encoding (e.g., ASCII, JPEG, MPEG-4).

By adding bits to each encoded unit, the redundancy allows both to detect errors in coded data and to correct them based on mathematical algorithms. Errors occur regularly in low probabilities due to random bit value flipping, or "physical bit fatigue," loss of the physical bit in storage its ability to maintain distinguishable value (0 or 1), or due to errors in inter or intra-computer communication. A random bit flip (e.g., due to random radiation) is typically corrected upon detection. A bit, or a group of malfunctioning physical bits (not always the specific defective bit is known; group definition depends on specific storage device) is typically automatically fenced-out, taken out of use by the device, and replaced with another functioning equivalent group in the device, where the corrected bit values are restored (if possible). The Cyclic redundancy check (CRC) method is typically used in storage for error detection.

Data compression

Data compression methods allow in many cases to represent a string of bits by a shorter bit string ("compress") and reconstruct the original string ("decompress") when needed. This allows to utilize substantially less storage (tens of percents) for many types of data at the cost of more computation (compress and decompress when needed). Analysis of trade-off between storage cost saving and costs of related computations and possible delays in data availability is done before deciding whether to keep certain data in a database compressed or not.

Data encryption

For security reasons certain types of data (e.g., credit-card information) may be kept encrypted in storage to prevent the possibility of unauthorized information reconstruction from chunks of storage snapshots (taken either via unforeseen vulnerabilities in a DBMS, or more likely, by bypassing it).

Data encryption is typically controlled through the DBMS's data definition interface, but in some cases may be a default and automatic.

Data storage types

This collection of bits describes both the contained database data and its related metadata (i.e., data that describes the contained data and allows computer programs to manipulate the database data correctly). The size of a database can nowadays be tens of Terabytes, where a byte is eight bits. The physical materialization of a bit can employ various existing technologies, while new and improved technologies are constantly under development. Common examples are:

- *Magnetic medium (e.g., in Magnetic disk)* - Orientation of magnetic field in magnetic regions on a surface of material (two orientation directions, for 0 and 1).
- *Dynamic random-access memory (DRAM)* - State of a miniature electronic circuit consisting of few transistors (among millions nowadays) in an integrated circuit (two states for 0 and 1).

These two examples are respectively for two major storage types:

- Nonvolatile storage can maintain its bit states (0s and 1s) without electrical power supply, or when power supply is interrupted;
- Volatile storage loses its bit values when power supply is interrupted (i.e., its content is erased).

Sophisticated storage units, which can, in fact, be effective dedicated parallel computers that support a large amount of nonvolatile storage, typically must include also components with volatile storage. Some such units employ batteries that can provide power for several hours in case of external power interruption and thus maintain the content of the volatile storage parts intact. Just before such a device's batteries lose their power the device typically automatically backs-up its volatile content portion (into nonvolatile) and shuts off to protect its data.

Databases are usually too expensive (in terms of importance and needed investment in resources, e.g., time, money, to build them) to be lost by a power interruption. Thus at any point in time most of their content resides in nonvolatile storage. Even if for operational reason very large portions of them reside in volatile storage (e.g., tens of Gigabytes in volatile memory, for in-memory databases), most of this is backed-up in nonvolatile storage. A relatively small portion of this, which temporarily may not have nonvolatile backup, can be reconstructed by proper automatic database recovery procedures after volatile storage content loss.

More examples of storage types:

- Volatile storage can be found in processors, computer memory (e.g., DRAM), etc.
- Non-volatile storage types include ROM, EPROM, Hard disk drives, Flash memory and drives, Storage arrays, etc.

Storage Metrics

Databases always use several types of storage when operational (and implied several when idle). Different types may significantly differ in their properties, and the optimal mix of storage types is determined by the types and quantities of operations that each storage type needs to perform, as well as considerations like physical space and energy consumption and dissipation (which may become critical for a large database). Storage types can be categorized by the following attributes:

- Volatile/Nonvolatile.
- Cost of the medium (e.g., per Megabyte), Cost to operate (cost of energy consumed per unit time).
- Access speed (e.g., bytes per second).
- Granularity — from fine to coarse (e.g., size in bytes of access operation).
- Reliability (the probability of spontaneous bit value change under various conditions).
- Maximal possible number of writes (of any specific bit or specific group of bits; could be constrained by the technology used (e.g., "write once" or "write twice"), or due to "physical bit fatigue," loss of ability to distinguish between the 0, 1 states due to many state changes (e.g., in Flash memory)).
- Power needed to operate (Energy per time; energy per byte accessed), Energy efficiency, Heat to dissipate.
- Packaging density (e.g., realistic number of bytes per volume unit)

Protecting storage device content: Device mirroring (replication) and RAID

While a group of bits malfunction may be resolved by error detection and correction mechanisms, storage device malfunction requires different solutions. The following solutions are commonly used and valid for most storage devices:

- *Device mirroring (replication)* - A common solution to the problem is constantly maintaining an identical copy of device content on another device (typically of a same type). The downside is that this doubles the storage, and both devices (copies) need to be updated simultaneously with some overhead and possibly some delays. The upside is possible concurrent read of a same data group by two independent processes, which increases performance. When one of the replicated devices is detected to be defective, the other copy is still operational, and is being utilized to generate a new copy on another device.
- *Redundant array of independent disks (RAID)* - This method generalizes the device mirroring above by allowing one device in a group of N devices to fail and be replaced with content restored (Device mirroring is RAID with N=2). RAID groups of N=5 or N=6 are

common. N>2 saves storage, when comparing with N=2, at the cost of more processing during both regular operation (with often reduced performance) and defective device replacement.

Device mirroring and typical RAID are designed to handle a single device failure in the RAID group of devices. However, if a second failure occurs before the RAID group is completely repaired from the first failure, then data can be lost. The probability of a single failure is typically small. Thus the probability of two failures in a same RAID group in time proximity is much smaller (approximately the probability squared, i.e., multiplied by itself). If a database cannot tolerate even such smaller probability of data loss, then the RAID group itself is replicated (mirrored). In many cases such mirroring is done geographically remotely, in a different storage array, to handle also recovery from disasters.

Database storage layout

Database bits are laid-out in storage in data-structures and grouping that can take advantage of both known effective algorithms to retrieve and manipulate them and the storage own properties. Typically the storage itself is design to meet requirements of various areas that extensively utilize storage, including databases. A DBMS in operation always simultaneously utilizes several storage types (e.g., memory, and external storage), with respective layout methods.

Database storage hierarchy

A database, while in operation, resides simultaneously in several types of storage. By the nature of contemporary computers most of the database part inside a computer that hosts the DBMS resides (partially replicated) in volatile storage. Data (pieces of the database) that are being processed/ manipulated reside inside a processor, possibly in processor's caches. These data are being read from/written to memory, typically through a computer bus (so far typically volatile storage components). Computer memory is communicating data (transferred to/from) external storage, typically through standard storage interfaces or networks (e.g., fibre channel, iSCSI). A storage array, a common external storage unit, typically has storage hierarchy of it own, from a fast cache, typically consisting of (volatile and fast) DRAM, which is connected (again via standard interfaces) to drives, possibly with different speeds, like flash drive and magnetic disk drives (non-volatile).

The drives may be connected to magnetic tapes, on which typically the least active parts of a large database may reside, or database backup generations.

Typically a correlation exists currently between storage speed and price, while the faster storage is typically volatile.

Data structures

A data structure is an abstract construct that embeds data in a well defined manner. An efficient data structure allows to manipulate the data in efficient ways. The data manipulation may include data insertion, deletion, updating and retrieval in various modes. A certain data structure type may be very effective in certain operations, and very ineffective in others. A data structure type is selected upon DBMS development to best meet the operations needed for the types of data it contains. Type of data structure selected for a certain task typically also takes into consideration the type of storage it resides in. In some DBMSs database administrators have the flexibility to select among options of data structures to contain user data for performance reasons. Sometimes the data structures have selectable parameters to tune the database performance.

Databases may store data in many data structure types. Common examples are the following:

- ordered/unordered flat files
- hash tables
- B+ trees
- ISAM
- heaps

Application data and DBMS data

A typical DBMS cannot store the data of the application it serves alone. In order to handle the application data the DBMS need to store this data in data structures that comprise specific data by themselves. In addition the DBMS needs its own data structures and many types of bookkeeping data like indexes and logs. The DBMS data is an integral part of the database and may comprise a substantial portion of it.

Database indexing

Indexing is a technique for improving database performance. The many types of indexes share the common property that they reduce the need to examine every entry when running a query. In large databases, this can reduce query time/cost by orders of magnitude. The simplest form of index is a sorted list of values that can be searched using a binary search with an adjacent reference to the location of the entry, analogous to the index in the back of a book. The same data can have multiple indexes (an employee database could be indexed by last name and hire date.)

Indexes affect performance, but not results. Database designers can add or remove indexes without changing application logic, reducing maintenance costs as the database grows and database usage evolves.

Given a particular query, the DBMS' query optimizer is responsible for devising the most efficient strategy for finding matching data.

Indexes can speed up data access, but they consume space in the database, and must be updated each time the data is altered. Indexes therefore can speed data access but slow data maintenance. These two properties determine whether a given index is worth the cost.

Database data clustering

In many cases substantial performance improvement is gained if different types of database objects that are usually utilized together are laid in storage in proximity, being clustered. This usually allows to retrieve needed related objects from storage in minimum number of input operations (each sometimes substantially time consuming). Even for in-memory databases clustering provides performance advantage due to common utilization of large caches for input-output operations in memory, with similar resulting behavior.For example it may be beneficial to cluster a record of an item in stock with all its respective order records. The decision of whether to cluster certain objects or not depends on the objects' utilization statistics, object sizes, caches sizes, storage types, etc.

Database materialized views

Often storage redundancy is employed to increase performance. A common example is storing materialized views, which are frequently needed External views. Storing such external views saves expensive computing of them each time they are needed.

Database and database object replication

Occasionally a database employs storage redundancy by database objects replication (with one or more copies) to increase data availability (both to improve performance of simultaneous multiple end-user accesses to a same database object, and to provide resiliency in a case of partial failure of a distributed database). Updates of a replicated object need to be synchronized across the object copies. In many cases the entire database is replicated.

DATABASE TRANSACTIONS

As with every software system, a DBMS that operates in a faulty computing environment is prone to failures of many kinds. A failure can corrupt the respective database unless special measures are taken to prevent this. A DBMS achieves certain levels of fault tolerance by encapsulating operations within transactions. The concept of a database transaction (or atomic transaction) has evolved in order to enable both a well understood database system behavior in a faulty environment where crashes can happen any time, and recovery from a crash to a well understood database state. A database transaction is a unit of work, typically encapsulating a number of operations over a database (e.g., reading a database object, writing, acquiring lock, etc.), an abstraction supported in database and also other systems. Each transaction has well defined boundaries in terms of which program/code executions are included in that transaction (determined by the transaction's programmer via special transaction commands).

ACID rules

Every database transaction obeys the following rules:

- *Atomicity* - Either the effects of all or none of its operations remain ("all or nothing" semantics) when a transaction is completed (committed or aborted respectively). In other words, to the outside world a committed transaction appears (by its effects on the database) to be indivisible, atomic, and an aborted transaction does not leave effects on the database at all, as if never existed.
- *Consistency* - Every transaction must leave the database in a consistent (correct) state, i.e., maintain the predetermined integrity rules of the database (constraints upon and among the database's objects). A transaction must transform a database from one consistent state to another consistent state (however, it is the responsibility of the transaction's programmer to make sure that the transaction itself is correct, i.e., performs correctly what it intends to perform (from the application's point of view) while the predefined integrity rules are

enforced by the DBMS). Thus since a database can be normally changed only by transactions, all the database's states are consistent. An aborted transaction does not change the database state it has started from, as if it never existed (atomicity above).

- *Isolation* - Transactions cannot interfere with each other (as an end result of their executions). Moreover, usually (depending on concurrency control method) the effects of an incomplete transaction are not even visible to another transaction. Providing isolation is the main goal of concurrency control.
- *Durability* - Effects of successful (committed) transactions must persist through crashes (typically by recording the transaction's effects and its commit event in a non-volatile memory).

Isolation, concurrency control, and locking

Isolation provides the ability for multiple users to operate on the database at the same time without corrupting the data.

Concurrency control comprises the underlying mechanisms in a DBMS which handle isolation and guarantee related correctness. It is heavily utilized by the Database and Storage engines both to guarantee the correct execution of concurrent transactions, and (different mechanisms) the correctness of other DBMS processes. The transaction-related mechanisms typically constrain the database data access operations' timing (transaction schedules) to certain orders characterized as the Serializability and Recoverability schedule properties. Constraining database access operation execution typically means reduced performance (rates of execution), and thus concurrency control mechanisms are typically designed to provide the best performance possible under the constraints. Often, when possible without harming correctness, the serializability property is compromised for better performance. However, recoverability cannot be compromised, since such typically results in a quick database integrity violation.

Locking is the most common transaction concurrency control method in DBMSs, used to provide both serializability and recoverability for correctness. In order to access a database object a transaction first needs to acquire a lock for this object. Depending on the access operation type (e.g., reading or writing an object) and on the lock type, acquiring the lock may be blocked and postponed, if another transaction is holding a lock for that object.

Query Optimization

A query is a request for information from a database. It can be as simple as "finding the address of a person with SS# 123-45-6789," or more complex like "finding the average salary of all the employed married men in California between the ages 30 to 39, that earn less than their wives." Queries results are generated by accessing relevant database data and manipulating it in a way that yields the requested information. Since database structures are complex, in most cases, and especially for not-very-simple queries, the needed data for a query can be collected from a database by accessing it in different ways, through different data-structures, and in different orders. Each different way typically requires different processing time. Processing times of a same query may have large variance, from a fraction of a second to hours, depending on the way selected. The purpose of query optimization, which is an automated process, is to find the way to process a given query in minimum time. The large possible variance in time justifies performing query optimization, though finding the exact optimal way to execute a query, among all possibilities, is typically very complex, time consuming by itself, may be too costly, and often practically impossible. Thus query optimization typically tries to approximate the optimum by comparing several common-sense alternatives to provide in a reasonable time a "good enough" plan which typically does not deviate much from the best possible result.

DBMS Support for Development of a Database

A DBMS typically intends to provide convenient environment to develop and later maintain an application built around its respective database type. A DBMS either provides such tools, or allows integration with such external tools. Examples for tools relate to database design, application programming, application program maintenance, database performance analysis and monitoring, database configuration monitoring, DBMS hardware configuration and related database mapping (especially for a distributed DBMS), storage allocation and database layout monitoring, storage migration, etc.

References

Childs, D. L. 1968, *Description of a set-theoretic data structure*, Technical Report 3 of the CONCOMP (Research in Conversational Use of Computers) Project, University of Michigan, Ann Arbor, Michigan, USA

Codd, E.F. (1970)."A Relational Model of Data for Large Shared Data Banks". In: *Communications of the ACM* 13 (6): 377–387.

itl.nist.gov (1993) *Integration Definition for Information Modeling (IDEFIX)*. 21 December 1993.

Ken North, "Sets, Data Models and Data Independence", *Dr. Dobb's*, 10 March 2010

Rainer, R. Kelly, and Casey G. Cegielski. Introduction to Information Systems: Supporting and Transforming Business. Hoboken, NJ: Wiley, 2012. Print.

William Hershey and Carol Easthope, "A set theoretic data structure and retrieval language", Spring Joint Computer Conference, May 1972 in *ACM SIGIR Forum*, Volume 7, Issue 4 (December 1972), pp. 45-55.

8

E-Commerce and E-Business

Electronic commerce or e-commerce refers to a wide range of online business activities for products and services. It also pertains to "any form of business transaction in which the parties interact electronically rather than by physical exchanges or direct physical contact."

E-commerce is usually associated with buying and selling over the Internet, or conducting any transaction involving the transfer of ownership or rights to use goods or services through a computer-mediated network. Though popular, this definition is not comprehensive enough to capture recent developments in this new and revolutionary business phenomenon. A more complete definition is: E-commerce is the use of electronic communications and digital information processing technology in business transactions to create, transform, and redefine relationships for value creation between or among organizations, and between organizations and individuals.

International Data Corp (IDC) estimates the value of global e-commerce in 2000 at US$350.38 billion. This is projected to climb to as high as US$3.14 trillion by 2004. IDC also predicts an increase in Asia's percentage share in worldwide e-commerce revenue from 5% in 2000 to 10% in 2004.

Electronic business, commonly referred to as "eBusiness" or "e-business", or an internet business, may be defined as the application of information and communication technologies (ICT) in support of all the activities of business. Commerce constitutes the exchange of products and services between businesses, groups and individuals and can be seen as one of the essential activities of any business. Electronic commerce focuses on the use of ICT to enable the external activities and relationships of the business with individuals, groups and other businesses.

The term "e-business" was coined by IBM's marketing and Internet teams in 1996.

Electronic business methods enable companies to link their internal and external data processing systems more efficiently and flexibly, to work more closely with suppliers and partners, and to better satisfy the needs and expectations of their customers.

In practice, e-business is more than just e-commerce. While e-business refers to more strategic focus with an emphasis on the functions that occur using electronic capabilities, e-commerce is a subset of an overall e-business strategy. E-commerce seeks to add revenue streams using the World Wide Web or the Internet to build and enhance relationships with clients and partners and to improve efficiency using the Empty Vessel strategy. Often, e-commerce involves the application of knowledge management systems.

E-business involves business processes spanning the entire value chain: electronic purchasing and supply chain management, processing orders electronically, handling customer service, and cooperating with business partners. Special technical standards for e-business facilitate the exchange of data between companies. E-business software solutions allow the integration of intra and inter firm business processes. E-business can be conducted using the Web, the Internet, intranets, extranets, or some combination of these.

Basically, electronic commerce (EC) is the process of buying, transferring, or exchanging products, services, and/or information via computer networks, including the internet. EC can also be beneficial from many perspectives including business process, service, learning, collaborative, community. EC is often confused with e-business.

E-COMMERCE VS E-BUSINESS

While some use e-commerce and e-business interchangeably, they are distinct concepts. In e-commerce, information and communications technology (ICT) is used in inter-business or inter-organizational transactions (transactions between and among firms/organizations) and in business-to-consumer transactions (transactions between firms/organizations and individuals).In e-business, on the other hand, ICT is used to enhance one's business. It includes any process that a business organization (either a for-profit, governmental or non-profit entity) conducts over a computer-mediated network. A more comprehensive definition of e-business is:

The transformation of an organization's processes to deliver additional customer value through the application of technologies, philosophies and computing paradigm of the new economy.

Three primary processes are enhanced in e-business:

1. *Production processes*, which include procurement, ordering and replenishment of stocks; processing of payments; electronic links with suppliers; and production control processes, among others;
2. *Customer-focused processes*, which include promotional and marketing efforts, selling over the Internet, processing of customers' purchase orders and payments, and customer support, among others; and
3. *Internal management processes*, which include employee services, training, internal information-sharing, video-conferencing, and recruiting. Electronic applications enhance information flow between production and sales forces to improve sales force productivity. Workgroup communications and electronic publishing of internal business information are likewise made more efficient.

Scope of E-Commerce

Electronic commerce is an emerging model of new selling and merchandising tools in which buyers are able to participate in all phases of a purchase decision, while stepping through those processes electronically rather than in a physical store or by phone (with a physical catalogue). The processes in electronic commerce include enabling a customer to access product information, select items to purchase, purchase items securely, and have the purchase settled financially.

It is an emerging concept that describes the process of buying and selling or exchanging of products, services; and information via computer networks including the Internet.

— From a communications]perspective, EC is the delivery of information, products/services, or payments over telephone lines, computer networks, or any other electronic means.

— From a business process perspective, EC is the application of technology toward the automation of business transactions and work flow.

— From a service perspective, EC is a tool that addresses the desire of firms, consumers, and management to cut service costs while improving the quality of goods and increasing the speed of service delivery.

— From an online perspective, EC provides the capability of buying and selling products and information on the Internet and other online services.

E-commerce is a term popularised by the advent of commercial services on the Internet. Internet e-Commerce is however, only one part of the overall sphere of e-commerce. The commercial use of the Internet is perhaps typified by once-off sales to consumers. Other types of transactions use other technologies. Electronic Markets (EMs) are in use in a number of trade segments with an emphasis on search facilities and Electronic Data Interchange (EDI) is used for regular and standardised transactions between organisations.

The mainstream of e-commerce consists of three areas: Electronic Markets, Electronic Data Interchange (EDI) and Internet Commerce these are represented as a diagram in Figure 1 and outlined in a little more detail below.

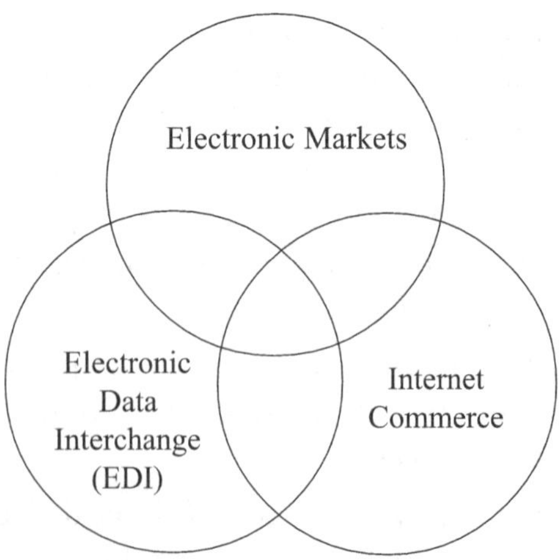

Figure 1. The three categories of E Commerce

Electronic Markets

An electronic market is the use of information and communications technology to present a range of offerings available in a market segment so that the purchaser can compare the prices (and other attributes) of the offerings and make a purchase decision. The usual example of an electronic market is an airline booking system.

Electronic markets are exampled by the airline booking systems. Electronic markets are also used in the financial and commodity markets and again the dealing is done via intermediaries; to buy stocks and shares a member of the public uses the services of a stockbroker. Arguably the use of electronic markets has served the customer well. With the assistance of a good travel agent the airline customer can be informed of all the flights available for an intended journey and then select, on the basis of price, convenience, loyalty scheme, etc. the flight that they wish to book.

Advantages and Disadvantages

The advantages of an electronic market to the customer are self-evident. Using an airline booking system, for example, there is a screen that shows all the flights from New York to Los Angeles and the consumer can make an informed choice without having to spend time and effort finding out which airlines fly that route and then contacting each of the airlines to obtain flight times, price and availability details. Once a flight is selected the system facilitates the booking of that flight, paying the fare and printing the ticket.

For the seller the advantages are less evident. The seller that is the most competitive may do well, the electronic market makes available information on their product and the advantage of that offering should be apparent. Less competitive suppliers are likely to be forced into price. Reductions and the competitive effect may force all suppliers to cut prices, possibly below the level at which it is possible to make a profit.

Electronic Data Interchange (EDI)

EDI provides a standardised system for coding trade transactions so that they can be communicated directly from one computer system to another without the need for printed orders and invoices and the delays and errors implicit in paper handling. EDI is used by organisations that make a large number of regular transactions. One sector where EDI is extensively used is the large supermarket chains, which use EDI for transactions with their suppliers.

Benefits of EDI

EDI can bring a number of advantages to the organisations that use it. It should save considerable time on the exchange of business transactions and has the potential for considerable savings in costs. EDI can be simply used to replace paper transactions with electronic transactions—this is the normal route taken in the initial installation of EDI. The full advantage of EDI is only realised when business practices are restructured to make full use of the potential of EDI; when EDI is used as an enabling technology to change the way the business operates-just-in-time (JIT) manufacture and quick response supply being prime examples of where EDI is used as an enabling technology to gain competitive advantage.

Shortened Ordering Time

Paper orders have to be printed, enveloped and sent out by the customer's post room, passed through the postal service, received by the supplier's post room, and input to the supplier's order processing system. To achieve all this, reliably, in under three days would be to do very well. EDI orders are sent straight into the network and the only delay is how often the supplier retrieves messages from the system. Orders can be in the supplier's system within a day, or if there is urgency the messages can be retrieved more frequency, for example every hour.

Cost Cutting

The use of EDI can cut costs. These include the costs of stationery and postage but these will probably be fully matched by the costs of running the EDI service. The principle saving from the use of EDI is the potential to save staff costs. The obvious example of this is that if the orders are directly input to the system there is no need for an order entry clerk. Note also that seasonal peak, staff holidays, etc. no longer create a backlog in the order entry area. The cost saving need to be offset against the system development and network casts.

Elimination of Errors

Keying any information into a computer system is a source of errors and keying paper orders into the order processing system is no exception. EDI eliminates this source of errors. On the down side, there is no order entry clerk who might have spotted errors made by the customer—the customer will get what the customer asked for.

Fast Response

With paper orders it would be several days before the customer was informed of any supply difficulty, such as the product is out of stock. With alternative product to be ordered or an alternative supplier to be used.

Accurate Invoicing

Just like orders, invoices can be sent electronically. EDI invoices have similar advantages to EDI orders in saved time and avoided errors. However, the major advantage in EDI invoices is that they can be automatically matched against the original order and cleared for payment without the sort of queries that arise when paper invoices are matched to orders.

EDI Payment

Payment can also be made by EDI. The EDI payment system can also generate an EDI payment advice that can be electronically matched against the relevant invoices, again avoiding query and delay. Indirect advantages of the use of EDI can be:

Reduced stock holding

The ability to order regularly and quickly reduces the amount of goods that need to be kept in a store room or warehouse at the shop or the factory. For many JIT manufacture and quick response supply systems stockholding is eliminated altogether with goods being delivered only as they are needed. Reduced stock holding cuts the cost of warehousing, the double handling goods (into store and then out again onto the factory or shop) and the capital requirement to pay for the goods that are just sitting in store.

Cash flow

Speeding up the trade cycle by getting invoices out quickly, and directly matched to the corresponding orders and deliveries, can and should speed up payments and hence improve cash flow. Elimination of most invoice queries can be particularly significant in reducing delays in payments.

Business opportunities

There is a steady increase in the number of customers, particularly large, powerful customers, that will only trade with suppliers that do business via EDI. Supermarkets and vehicle assemblers are prime examples. Being ready and able to trade electronically can be an advantage when competing for new business.

Customer lock-in

An established EDI system should be of considerable advantage to both customer and supplier. Switching to a new supplier requires that the electronic trading system and trading relationship be redeveloped, a problem to be avoided if a switch of supplier is not essential.

To gain these advantages EDI has to be seen as an investment there are costs upfront and the payback is longer term. The costs is the set up of the EDI system (hardware, software and network) and the time required to establish agreements with trading partners. The savings only start when there is a significant volume of business transacted using EDI, a point that is called the 'critical mass' in the jargon of EDI.

Internet Commerce

Information and communications technologies can also be used to advertise and make once-off sales of a wide range of goods and services. This type of e-Commerce is typified by the commercial use of the Internet. The Internet can, for example, be used for the purchase of books that are then delivered by post or the booking of tickets that can be picked up by the clients when they arrive at the event. It is to be noted that the Internet is not the only technology used for this type of service and this is not the only use of the Internet in e-Commerce

E-COMMERCE FRAMEWORK

From the business activity already taking place, it is clear that E-commerce applications will be built on the existing technology infrastructure-a myriad of computers, communications networks, and communication software forming the nascent Information Superhighway. Figure 2 shows a variety of possible e-commerce applications; including both inter organisational and consumer oriented examples. None of these uses would be possible without each of the building blocks in the infrastructure:

- Common business services, for facilitating the buying and selling process
- Messaging and information distribution, as a means of sending and retrieving information
- Multimedia content and network publishing, for creating a product and a means to communicate about it

— The Information Superhighway-the very foundation-for providing the highway system along which all e-commerce must travel the two pillars supporting all e-commerce applications and infrastructure-are just as indispensable.

— Public policy, to govern such issues as universal access, privacy, and information pricing

— Technical standards, to dictate the nature of information publishing, user interfaces, and transport in the interest of compatibility across the entire network.

To better understand the integration of the various infrastructure components in our framework, let us use the analogy of a traditional transportation business. Any successful e-commerce application will require the I-way infrastructure in the same way that regular commerce needs the interstate highway network to carry goods from point to point. You must travel across this highway, whether you are an organisation purchasing supplies or a consumer ordering a movie on demand. Understand, however, that the I-way is not one monolithic data highway designed according to long-standing, well defined rules and regulations based on well-known needs. Rather, still under construction, the I-way will be a mesh of interconnected data highways of many forms: telephone wires, cable TV wires, radio-based wireless-cellular and satellite.

Far from complete, the I-way is quickly acquiring new on-ramps and even small highway systems. The numerous constructors are either in competition with or in alliance with one another, all in an effort to convince traffic to use their on-ramps or sections of the highway because, like toll ways, revenues in E-commerce are based on vehicular traffic, in the case, vehicles transporting information or multimedia content. The myriad transactions among businesses means that the ultimate winner must select the technology for the I-way that best matches future business needs by using today's tools. Building an access road to a ghost town or a highway too narrow to handle the traffic will yield equally little return on investment for those who have been less successful at matching needs with the infrastructure. Building the various highways is not enough. Transport vehicles are needed, routing issues must be addressed, and of course the transportation costs must be paid. On the I-way, the nature of vehicular traffic is extremely important. The information and multimedia content determines what type of vehicle is needed. A breakdown of potential everyday e-commerce vehicles into their technological components shows that they vary widely in complexity and may even need to travel different routes on the I-way, much the way an eighteen-wheeler may be restricted from travelling roads that cannot accommodate it:

Movies = video + audio

Digital games = music + video + software

Electronic books = text + data + graphics + music + photographs + video.

Once these vehicles (multimedia content) are created, where are they housed? What sort of distribution warehouses is needed to store and deliver their multimedia cargo? In the electronic "highway system" multimedia content is stored in the form of electronic documents. These documents are often digitised, compressed, and stored in computerised libraries or multimedia storage warehouses called servers that are linked by transport networks to each other and to the software/hardware clients that allow customers to access them. Exactly how do the vehicles move from one distribution warehouse to another? In a traditional transportation business, diesel engines or gasoline powered motors move the trucks along the roadways.

Standards are crucial in the world of global e-commerce, to ensure not only seamless and harmonious integration across the transportation network but access of information on any type of device the consumer choose-laser disc, PCs, portable hand-held devices or television + set-top boxes (cable converter boxes)-and on all types of operating systems. For example, without the adoption of video standards, video conferencing will never become widespread, as each manufacturer will attempt to develop equipment that maximises their short-term profits rather than working toward customer goals such as interoperability. The concept of "convergence" is essential to the operation of the Information Superhighway and to the way the business world is gearing up to deal with it.

Architectural Framework

The software framework necessary for building electronic commerce applications is little understood in existing literature. In general a framework is intended to define and create tools that integrate the information found in today's closed systems and allow the development of e-commerce applications. It is important to understand that the aim of the architectural framework itself is not to build new database management systems, data repository, computer languages, software agent based transaction monitors, or communication protocols. Rather, the architecture should focus on synthesising the diverse resources already in place in corporations to facilitate the integration of data and software for better applications. The electronic commerce application architecture consists of six layers of functionality, or services:

(1) applications;

(2) brokerage services, data or transaction management;

(3) interface, and; support layers

(4) secure messaging, security and electronic document interchange;

(5) middle ware and structured document interchange; and

(6) network infrastructure and basic communications services.

These layers cooperate to provide a seamless transition between today's computing resources and those of tomorrow by transparently integrating information access and exchange within the context of the chosen application. Electronic commerce applications are based on several elegant technologies. But only when they are integrated do they provide uniquely powerful solutions. In the ensuing discussion of each of these layers, we will not elaborate on the various aspects of the network infrastructure that transports information.

Application Services of EC

The application services layer of e-commerce will be comprised of existing and future applications built on the innate architecture. Three distinct classes of electronic commerce application can be distinguished: customer to business, business-to-business, and intra organisation.

Consumer-to-Business Transactions

This category also calls marketplace transaction. In a marketplace transaction, customers learn about products differently through electronic publishing, buy them differently using electronic cash and secure payment systems, and have them delivered differently. Also, how customers allocate their loyalty may also be different. In light of this, the organisation itself has to adapt to a world where the traditional concepts of brand differentiation no longer holdwhere "quality" has a new meaning, where "content" may not be equated to "product," Where "distribution" may not automatically mean "physical transport." In this new environment, brand equity can rapidly evaporate forcing firms to develop new ways of doing business.

Business-to Business Transactions

This category also call market-link transaction. Here, businesses, governments, and other organisations depend on computer -to computer communication as a fast, an economical, and a dependable way to conduct business' transactions. Small companies are also beginning to see the benefits of adopting the same methods. Business-to-business transactions include the use of EDI and electronic mail for purchasing goods and services, buying information and consulting services, submitting requests for proposals, and receiving proposals. Examine this scenario.

The current accounts payable process occurs through the exchange of paper documents. Each year the trading partners exchange millions of invoices, checks, purchase orders, financial reports, and other transactions. Most of the documents are in electronic form at their point of origin but are printed and key-entered at the point of receipt. The current manual process of printing, mailing is costly, time consuming, and error-prone. Given this situation and faced with the need to reduce costs, small businesses are looking toward electronic commerce as a possible saviour.

Intra-organisational Transactions

This category also call market-driven transactions. A company becomes market driven by dispersing throughout the firm information about its customers and competitors; by spreading strategic and tactical decision making so that all units can participate; and by continuously monitoring their customer commitment by making improved customer satisfaction an ongoing objective. To maintain the relationships that are critical to delivering superior customer value, management must pay close attention to service, both before and after sales. In essence, a market-driven business develops a comprehensive understanding of its customers' business and how customers in the immediate and downstream markets perceive value.

Three major components of market-driven transactions are

— customer orientation through product and service
— customisation; cross-functional coordination through enterprise
— integration; and advertising, marketing, and customer service.

Information Brokerage and Management

The information brokerage and management layer provides service integration through the notion of information brokerages, the development of which is necessitated by the increasing information resource fragmentation. We use the notion of information brokerage to represent an intermediary who provides service integration between customers and information providers, given some constraint such as a low price, fast service, or profit maximisation for a client. Information brokers, for example, are rapidly becoming necessary in dealing with the voluminous amounts of information on the networks.

As on-line databases migrate to consumer information utilities, consumers and information professionals will have to keep up with the knowledge, and ownership/of all these systems. Who's got what? How do you use it? What do they charge? Most professionals have enough trouble keeping track of files of 1 interest on one or two database services. With all the complexity associated with large numbers of on-line databases and service bureaus, if it is impossible to expect humans to do the searching. It will have to be software programs information brokers or software agents, to use the more popular term-that act on the searcher's behalf.

Information brokerage does more than just searching. It addresses the issue of adding value to the information that is retrieved. For instance, in foreign exchange trading, information is retrieved about the latest currency exchange rates in order to hedge currency holdings to minimise risk and maximise profit. In other words, the act of retrieving the information is the input to other transactions. With multiple transactions being the norm in the real world, service integration becomes critical.

Taking the same foreign exchange example further, service integration allows one to link the hedging program (offered on a time-sharing basis by a third party) with the search program (could be another vendor) that finds the currency rates from the cheapest on-line service to automatically send trades to the bank or financial services company. In effect, a personalised automated trading system can be created without having to go to any financial institution. This is just one example of how information brokerages can add value.

Another aspect of the brokerage function is the support for data management and traditional transaction services. Brokerages may provide tools to accomplish more sophisticated, time-delayed updates or future compensating transactions. These tools include software agents, distributed query generator, the distributed transaction generator, and the declarative resource constraint base which describes a business's rules and-environment information. At the heart of this layer lies the work-flow scripting environment built on a software agent model that coordinates work and data flow among support services. As pointed out earlier, software agents are used to implement information brokerages.

Software agents are mobile programs that have been called "healthy viruses," "digital butlers/" and "intelligent agents." Agents are encapsulations of users' instruction that perform all kinds of tasks in electronic marketplaces spread across networks. Information brokerages dispatch agents capable of information resource gathering, negotiating deals, and performing transactions. The agents are intelligent because they have contingency plans of action. They examine themselves and their environment and if necessary change from their original course of action to an alternative plan.

For example, suppose you send an agent to an on-line store with a request to order a bouquet of roses for $25 or less. If the shop offers roses starting at $30, your agent can either choose a different bouquet or find a different store by consulting an on-line "Yellow Pages" directory, depending on prior instructions. Although the notion of software agents sounds very seductive, it will take a while to solve the problems of interregna communication, interoperable agents, and other headaches that come with distributed computing and networking. To some critics, the prospect of a single-agent language like Telescript as a world standard is disturbing. They worry that agents sound a bit too much like computer viruses, which instead of running errands may run amok. Vendors such as General Magic go to great lengths to explain the precautions it has taken to make this impossible: the limits placed on the power of agents, the "self-destruct" mechanism built into their codes. Yet until electronic commerce services are up and running on a large scale, it is impossible to know how well software agents will work.

Interface and Support Services

The third layer, interface and support services, will provide interfaces for electronic commerce applications such as interactive catalogues and will support directory services-functions necessary for information search and access. These two concepts are very different. Interactive catalogues are the customised interface to consumer applications such as home shopping. An interactive catalogue is an extension of the paper-based catalogue and incorporates additional features such as sophisticated graphics and video to make the advertising more attractive.

Directories, on the other hand, operate behind the scenes and attempt to organise the enormous amount of information and transactions generated to facilitate electronic commerce. Directory services databases make data from any server appear as a local file. A classic example of a directory is the telephone White Pages, which allows us to locate people and telephone numbers. In the case of electronic commerce, directories would play an important role in information management functions. For instance, take the case of buying an airline ticket with several stopovers with the caveat that the time between layovers be minimised. This search would require several queries to various on-line directories to-find empty seats on various airlines and then the availability of seats would; be coordinated with the amount of time spent in the airport terminals.

The primary difference between the two is that unlike interactive catalogues, which deal with people, directory support services interact directly with software applications. For this reason, they need not have the multimedia glitter and jazz generally associated with interactive catalogues. From a computing perspective, we can expect that there will be no one common user interface that will glaze the surface of all electronic commerce applications, but graphics and object manipulation will definitely predominated. Tool developers and designers might incorporate common tools for interface building, but the shape of catalogues or directories will depend on the users' desires and functional requirements.

Secure Messaging and Structured Document Interchange Services

The importance of the fourth layer, secured messaging, is clear. Everyone in business knows that electronic messaging is a critical business issue. Consider a familiar business scenario:

You hand over an urgent fax Monday and find out Tuesday that it's still sitting on your fax operator's desk. What happened? The line was busy and he thought he'd try again later. Or, the number was wrong, but he forgot to let you know. Or you're in London and you need to send a spreadsheet that details a marketing plan for a product introduction strategy to a co-worker in New York. This must be done today, not tomorrow when the courier service would deliver. There is a solution to these common and frustrating problems. It's called Integrated Messaging: a group of computer services that through the use of a network send, receive, and combine messages, faxes, and large data files. Some better-known examples are electronic mail, enhanced fax, and electronic data interchange.

Broadly defined, messaging is the software that sits between the network infrastructure and the clients or electronic commerce applications, masking the peculiarities of the environment. Others define messaging as a framework for the total implementation of portable applications, divorcing you from the architectural primitives of your system. In general, messaging products are not applications that solve problems; they are more enablers of the applications that solve problems. Messaging services offer solutions for communicating non formatted (unstructured) data-letters, memos, reports as weft as formatted (structured) data such as purchase orders, shipping notices, and invoices.

Unstructured messaging consists of fax, e-mail, and form-based systems like Lotus Notes. Structured documents messaging consist of the automated interchange of standardised and approved messages between computer applications, via telecommunications lines. Examples of structured document messaging include EDI. Messaging is gaining momentum in electronic commerce and seems to have many advantages. It supports both synchronous (immediate) and asynchronous (delayed) message delivery and processing. With asynchronous messaging, when a message is sent, work continues (software doesn't wait for a response). This allows the transfer of messages through store-and-forward methods.

Another advantage of messaging is that it is not associated with any particular communication protocol. No preprocessing is necessary, although there is an increasing need for programs to interpret the message. Messaging is well suited for both client server and peer-to-peer computing models. In distributed systems, the messages are treated as "objects" that pass between systems. Messaging is central to work-group computing that is changing the way businesses operate. The ability to access the right information at the right time across diverse work groups is a challenge.

Today, with the messaging tools, people can communicate and work together more effectively-no matter where they are located. When an employee sends an electronic mail form, the information travels along with the form. So one person can start the form, mail it to the next person, fill it in/ sign it, mail it to the next, and so on. This is known as message-enabled workflow solutions.

The main disadvantages of messaging are the new types of applications it enables-which appear to be more complex, especially to traditional programmers-and the jungle of standards it involves. Because of the lack of standards, there is often no interoperability between different messaging vendors leading to islands of messaging. Also, security, privacy, and confidentiality through data encryption and authentication techniques are important issues that need to be resolved for ensuring the legality of the message-based transactions themselves.

Middleware Services

Middleware is a relatively new concept that emerged only recently. Like so many other innovations, it came into being out of necessity. Users in the 1970s, when vendors, delivered homogeneous systems that worked, didn't have a need for middleware conditions changed-along with the hardware and the software the organisations couldn't cope: The tools were inadequate, the backlog was enormous, and the pressure was overwhelming. And, the users were dissatisfied. Something was needed to solve all the interface, translation, transformation, and interpretation problems that were driving application developers crazy.

With the growth of networks, client-server technology, and all other forms of communicating between/among unlike platforms, the problems of getting all the pieces to work together grew from formidable to horrendous as the cry for distributed computing spread, users demanded interaction between dissimilar systems, networks that permitted shared resources, and applications that could be accessed by multiple software programs. In simple terms, middleware is the ultimate mediator between diverse software programs that enables them talk to one another. Another reason for middleware is the computing shift from application centric to data centric. That is, remote data controls all of the applications in the network instead of applications controlling data. To achieve data-centric computing, middleware services focus on three elements:

— transparency,

— transaction security and management, and

— distributed object management and services.

Transparency

Transparency implies that users should be unaware that they are accessing multiple systems. Transparency is essential for dealing with higher-level issues than physical media and interconnection that the underlying network infrastructure is in charge of. The ideal picture is one of a "virtual" network: a collection of workgroup, departmental, enterprise, and inter enterprise LANs that appears to the end user or client application to be a seamless and easily accessed whole.

Transparency is accomplished using middleware that facilitates a distributed computing environment. This gives users and applications transparent access to data, computation, and other resources across collections of multi-vendor, heterogeneous systems. The strategic architectures of every major system vendor are now based on some form of middleware. The key to realising the theoretical benefit of such architecture is transparency. Users need not spend their time trying to understand where something is. Nor should application developers have to code into their applications the exact locations of resources over the network. The goal is for the applications to send a request to the middleware layer, which then satisfies the request any way it can, using remote information.

Transaction security and management

Support for transaction processing (TP) is fundamental to success in the electronic commerce market. Security and management are essential to all layers in the electronic commerce model. At the transaction security level, two broad general categories of security services exist: authentication and authorisation. Transaction integrity must be a given for businesses that cannot afford any loss or inconsistency in data. Some commercial sites have had gigantic centralised TP systems running for years. For electronic commerce, middleware provides the qualities expected in a standard TP system: the so-called ACID properties (atomicity, consistency, isolation, and durability).

Distributed object management and services

Object orientation is proving fundamental to the proliferation of network-based applications for the following reasons: It is too hard to write a network-based application without either extensive developer retraining or a technology that camouflages the intricacies of the network. Objects are defined as the combination of data and instructions acting on the data. Objects are an evolution of the more traditional programming concept of functions and procedures. A natural instance of an object in electronic commerce is a document. A document carries data and often carries instructions about the actions to be performed on the data.

Today, the term object is being used interchangeably with document resulting in a new form of computing called document oriented computing. Here, the trend is to move away from single data-type documents such as text, pictures, or video toward integrated documents known as compound document architectures. The best example of this approach is an active document. If you create a new document that is an integration of the spreadsheet, word processor, and presentation package, what you'll see in the next generation of operating systems is that as you scroll through your document, the tool bar will automatically change from a spreadsheet too bar, to a word processing tool bar, to a presentation package tool bar. These applications will also be able to access and retrieve data from any file in the computing network.

Consumer Oriented E-commerce

E-retailing essentially consists of the sale of goods and services. We can divide tangible goods into two categories: physical goods and digital goods. Examples of physical goods would be a book, a television set, a video recorder, a washing machine, etc. Examples of digital goods are software and music, which may be downloaded from the internet. The sale of intangible goods is sometimes called e-servicing. Examples of services that may be sold are information such as the most recent stock prices, the most recent foreign exchange rate, or education.

Entertainment such as -games that would be played on the internet are also examples of e-services. So are the sales of services such as telecommunication services or banking services. The sale of tangible and intangible goods are all referred to as customer oriented e-commerce or eretailing, if they are sold directly to the consumer who is the end user.

Traditional Retailing

Traditional retailing essentially involves selling to a final customer through a physical outlet or through direct physical communication. This normally involves a fairly extensive chain starting from a manufacturer to a wholesaler and then to the retailer who through a physical outlet has direct contact with the final customer. Examples of physical outlets that retailers currently use are:

— malls
— generalised stores (e.g. department store)
— specialised stores
— franchise stores

It is useful to reflect that even in traditional retailing we have moved away from just using a static physical outlet within which a customer can have direct contact with the retailer. Thus, more recent forms of traditional retailing include

— direct mailing
— telemarketing
— door-to-door sales
— vending machines

Direct mailing to a customer normally involves sending a brochure or catalogue to a customer. The customer browses through this catalogue and then carries out mail ordering. In some respects, this notion of browsing through a catalogue is a forerunner of e-retailing. Direct mailing, telemarketing, door-to-door sales, or the use of vending machines includes other forms that have actually moved away from a physical fixed outlet and in a way are an intermediate form of the movement away from traditional physical retailing outlet to the virtual retailing we see on the internet.

E-Retailing

The internet has allowed a new kind of specialisation to emerge. Instead of specialising just in a special product line, they allow specialisation in particular classes of customers and sellers. Emails essentially provide a web-hosting service for your individual store much in the way that mall provide a hosting service in the sense of a physical location for your store.

In the future we may see the equivalent of franchise stores developing. One new class of business that is developing very quickly on the internet is the e-broker. The e-broker does not sell directly to a customer but brings the customer in touch with a particular supplier, so that a given set of criteria specified by the customer is satisfied. For example, the customer may want to buy goods at the cheapest price and so the e-broker would then do a search to find the supplier that would provide the cheapest goods. Or, a customer may want to find a particular kind of goods and the e-broker sets about determining which supplier would provide those goods. This area of e-broking is likely to grow very greatly in the near future.

Advantages of E-Retailing

Customers enjoy a number of benefits from e-retailing. The first of these is convenience. It is convenient for the customer as he does not have to move from shop to shop physically in order to examine goods. He is able to sit in front of a terminal and search the net and examine the information on goods. The second aspect of convenience he gets is in terms of time.

Normally, the traditional shop has an opening time and a closing time and the customer can only visit the shop within these periods. On the net, the customer can choose at any time to visit a site to examine the goods that are available and actually carry out his purchasing at one's own convenient time. The third type of convenience that the customer gets is that he has access to a search engine, which will actually locate the products that he describes' and also the site where they may be available, or perhaps even locate the sites where they may be available at the best price The second type of benefit to customers is better information.

The Internet and the World Wide web are essentially communication media that allow retailers to put on quite extensive information related to their products, which is available to the customers. Furthermore, since the customer can look at several sites, he will be able to obtain different pieces of information from each site to build a far better picture for himself about the products that he is interested in.

In some sites, there are customer reviews of different products as well as reviews by the business itself. An example of this can be found on Amazon.com. This allows-the customer to finesse his requirements before actually making the purchase. It also gives different sources of information. The third type of benefit that the customer gets is competitive pricing. This is due to two factors.

— The first is lowered costs to the retailer because he does not have to maintain a physicals how room, he does not have to hire several shop assistants, and these savings can be passed on to customers in the form of reduced prices.

— Secondly, competitive pricing pressure that arises from the fact that the customer is now able to look at prices at several sites. Therefore, the pressure is always there on the retailer to maintain a competitive price for his products.

— The third benefit is customisation. The customer can actually specify the features of the products that he would like and thus in some cases it is possible that the retailer may allow a customised product to be delivered.

An example of this is on the Dell site. The computer site allows shoppers to custom specify their own computer software and hardware configurations. Thus, the customer is able to select exactly what he wants. This ability to get the business to deliver a product that the customer specifies he wants is the essence of C2B E-commerce. In summary, the benefits of e-retailing to the customer include

— convenience

— better information

— competitive price .

— customisation

— shopping anywhere, anytime

So with e-retailing, the customer can shop "anywhere around the globe without being restricted to his local vicinity. He could, for example, purchase goods over_ and have them delivered to a domestic address. He can also shop, as mentioned earlier at any time. These are very considerable benefits of e-retailing to the customer. These benefits could see larger and larger numbers of customers move more and more of their shopping on to eretailing sites in the future.

There are a number of benefits of e-retailing to the business itself.

— The first of these is global reach. The retailer now is no longer restricted to customers who are able to reach the store physically. They can be from anywhere around the globe. The retailer must, of course, deliver the goods of a purchase to the customer.

— The second benefit is better customer service. The use of email and the use of electronic interchange of messages between the customer and the retailer allows better communication between the customer and the retailer. These allow one to easily inquiries and deal with complaints. These also allow a much more rapid response time than was possible in the days of faxes and postal mail.

— The third benefit is the lowered capital cost to the retailer. The retailer does not have to maintain showrooms, he can probably have lower inventories. Thus, while Amazon.com lists over a few million titles, it keeps an inventory of a few thousand best selling titles only. Therefore, the retailer has lower warehousing costs. He does not have to have many shop assistants who are physically answering questions and. Showing the customer goods.

— The fourth benefit to the retailer is mass customisation. Based on requests by the customers, the retailer is now able to carry out mass customisation with reduced time to market for the customised products.

— The next advantage is targeted marketing. The retailer is now able to pick on a specific targeted group of customers and direct marketing towards these customers. The retailer is also able to provide more value-added services in the way of better information, add-on services to basic services, or add-on options to products that he is selling.

— The last advantage to the retailer consists of different new forms of specialised stores that he is now able to utilise.

It also creates new opportunities for niche marketing. The benefits to the e-retailer are:

— global reach

— better customer service

— low capital cost

— mass customisation

— targeted marketing

— more value-added services

— new forms of specialised stores and niche marketing

Success factors for traditional retailing

There are a number of key success factors which have been identified for traditional retailing . Two of these are the size and the number of outlets. The larger the retailer, the greater the buying muscle and therefore the lower the price for procurement. The number of outlets also allows the retailer to spread the purchase costs over a larger inventory. In addition, the number of outlets provides for better visibility. The retailer is now visible to the customer at many geographical locations rather than just one. Location is, of course, an extremely important success factor in traditional retailing. The retailer may choose to be sited in the central business district, in a regional area, in a shopping complex, or in a street of shops. This may relate to the category of customers and the costs associated with the site.

Other factors that are very important in traditional retailing are store atmosphere and store layout Store atmosphere evokes a particular look and feel about the retailer and is therefore important to the positioning in the market. Store layout is important in creating an atmosphere but is also important in ensuring that one groups different sets of products together, so that the purchase of one product will frequently lead to the purchase of another, thus allowing for cross selling. Price is important and here it is not necessarily the cheapest price, but the price which is consonant with what the customer expects to pay for the goods.

The variety of goods in the case of a large store, particularly of a department store, is also important because a customer would come in looking for one set of goods and then choose to purchase others. Profit margins are important in traditional retailing, and last but not least is the level of turnover. To summarise; the key success factors for traditional retailing are:

— size
— number of outlets
— visibility
— location
— store atmosphere
— store layout
— price
— variety of goods
— profit margins
— turnover

The first one considered is size. The important point to realise here is that no matter how large the company is your e-store presentation is still-limited to the size of the computer screen, which may be, say, 15 in. It is not necessary to look at the number of outlets in cyberspace because you probably need only one web set up. However, this web site is linked to other similar web sites and portals. Therefore, it is not the number of outlets that is important but the number of links from other important sites to your web site which is far more important. When one thinks of visibility, it is all the more important in cyberspace. It is notes enough to create a web site; you have to let the world know the existence of your 'web site and that people can purchase from you're-store.

When looking at visibility, important point to realise here is that most people find information on the Internet: trough the use of search engines. Therefore, it is very important to register the web or e-store with the most common or the most widely used search engines, such LS-Lycos, Alta Vista, and Yahoo. It is also important to link your web site with other well-known web sites that have similar interests, or major portals such as Yahoo, which link back to your e-store. This can greatly increase the visibility of the web site.

When considering location, we note that the geographic boundaries no longer exist. A local e-store and a foreign e-store are both just "one click" away. Store atmosphere is particularly important on the web. The "look and feel" of web site should match with the company's image as well as the market position Look it seeks to address. Thus, if you are selling very up-market clothes such as Gucci and Armani, your web site for these stores should have a sophisticated look and feel rout it. On the other hand, if you are selling other kinds of goods, you could choose to have a slightly jazzier image on your web site.

The front page or the homepage of your e-store is particularly important. It may be the only chance that you get of. Wring a potential customer into your e-store. In some ways it plays a similar role fur the e-store that the window display plays for the traditional store. What is also important is in going through this store. The layout of the store has to be such that it facilitates the customer's interests. The advantage of using store layouts in the e-store is that the layout can actually be made dynamic and be determined by the customer's interests.

The customer's interests could be obtained from data mining his previous purchases at the e-store. This is the major difference between a traditional store and an e-store. Price is very important in cyberspace because the customer can easily carry out comparison pricing between your e-store and other e-stores not just in your immediate neighbourhood but all around the world. Also some e-brokers provide agents or services that carry out comparison pricing; therefore, the customer can easily find me cheapest price. For this reason, it is important that in eretailing one sets up a competitive pricing structure.

Next, when one looks at the variety of goods, one here needs to consider very carefully whether you are setting up a specialised e-store or an e-department store. If you are setting up a specialised e-store, then you need to gain access to the specific target group of customers you are interested in because they can travel so easily in cyberspace to reach you. When one looks at profit margin and turnover, generally profit margins per item tend to be lower with e-retailing, and so turnover must be higher.

Lastly, if one examines profitability one finds that this is still important with the e-stores, but in recent times the share price of the e-store appears to have assumed more importance in the eyes of investors. However, in the long run, profitability will assume more importance.

Electronic Business Security

E-Business systems naturally have greater security risks than traditional business systems, therefore it is important for e-business systems to be fully protected against these risks. A far greater number of people have access to e-businesses through the internet than would have access to a traditional business.

Customers, suppliers, employees, and numerous other people use any particular e-business system daily and expect their confidential information to stay secure. Hackers are one of the great threats to the security of e-businesses. Some common security concerns for e-Businesses include keeping business and customer information private and confidential, authenticity of data, and data integrity.

Some of the methods of protecting e-business security and keeping information secure include physical security measures as well as data storage, data transmission, anti-virus software, firewalls, and encryption to list a few.

Privacy and Confidentiality

Confidentiality is the extent to which businesses makes personal information available to other businesses and individuals. With any business, confidential information must remain secure and only be accessible to the intended recipient. However, this becomes even more difficult when dealing with e-businesses specifically. To keep such information secure means protecting any electronic records and files from unauthorized access, as well as ensuring safe transmission and data storage of such information. Tools such as encryption and firewalls manage this specific concern within e-business.

Authenticity

E-business transactions pose greater challenges for establishing authenticity due to the ease with which electronic information may be altered and copied. Both parties in an e-business transaction want to have the assurance that the other party is who they claim to be, especially when a customer places an order and then submits a payment electronically. One common way to ensure this is to limit access to a network or trusted parties by using a virtual private network (VPN) technology. The establishment of authenticity is even greater when a combination of techniques are used, and such techniques involve checking "something you know" (i.e. password or PIN), "something you need " (i.e. credit card), or "something you are" (i.e. digital signatures or voice recognition methods). Many times in e-business, however, "something you are" is pretty strongly verified by checking the purchaser's "something you have" (i.e. credit card) and "something you know" (i.e. card number).

Data Integrity

Data integrity answers the question "Can the information be changed or corrupted in any way?" This leads to the assurance that the message received is identical to the message sent. A business needs to be confident that data is not changed in transit, whether deliberately or by accident. To help with data integrity, firewalls protect stored data against unauthorized access, while simply backing up data allows recovery should the data or equipment be damaged.

Non-repudiation

This concern deals with the existence of proof in a transaction. A business must have assurance that the receiving party or purchaser cannot deny that a transaction has occurred, and this means having sufficient evidence to prove the transaction. One way to address non-repudiation is using digital signatures. A digital signature not only ensures that a message or document has been electronically signed by the person, but since a digital signature can only be created by one person, it also ensures that this person cannot later deny that they provided their signature.

Access Control

When certain electronic resources and information is limited to only a few authorized individuals, a business and its customers must have the assurance that no one else can access the systems or information. Fortunately, there are a variety of techniques to address this concern including firewalls, access privileges, user identification and authentication techniques (such as passwords and digital certificates), Virtual Private Networks (VPN), and much more.

Availability

This concern is specifically pertinent to a business' customers as certain information must be available when customers need it. Messages must be delivered in a reliable and timely fashion, and information must be stored and retrieved as required. Because availability of service is important for all e-business websites, steps must be taken to prevent disruption of service by events such as power outages and damage to physical infrastructure. Examples to address this include data backup, fire-suppression systems, Uninterrupted Power Supply (UPS) systems, virus protection, as well as making sure that there is sufficient capacity to handle the demands posed by heavy network traffic.

Common Security Measures

Many different forms of security exist for e-businesses. Some general security guidelines include areas in physical security, data storage, data transmission, application development, and system administration.

Physical security

Despite e-business being business done online, there are still physical security measures that can be taken to protect the business as a whole. Even though business is done online, the building that houses the servers and computers must be protected and have limited access to employees and other persons. For example, this room should only allow authorized users to enter, and should ensure that "windows, dropped ceilings, large air ducts, and raised floors" do not allow easy access to unauthorized persons. Preferably these important items would be kept in an air-conditioned room without any windows.

Protecting against the environment is equally important in physical security as protecting against unauthorized users. The room may protect the equipment against flooding by keeping all equipment raised off of the floor. In addition, the room should contain a fire extinguisher in case of fire. The organization should have a fire plan in case this situation arises.

In addition to keeping the servers and computers safe, physical security of confidential information is important. This includes client information such as credit card numbers, checks, phone numbers, etc. It also includes any of the organization's private information. Locking physical and electronic copies of this data in a drawer or cabinet is one additional measure of security. Doors and windows leading into this area should also be securely locked. Only employees that need to use this information as part of their job should be given keys.

Important information can also be kept secure by keeping backups of files and updating them on a regular basis. It is best to keep these backups in a separate secure location in case there is a natural disaster or breach of security at the main location.

"Failover sites" can be built in case there is a problem with the main location. This site should be just like the main location in terms of hardware, software, and security features. This site can be used in case of fire or natural disaster at the original site. It is also important to test the "failover site" to ensure it will actually work if the need arises.

State of the art security systems, such as the one used at Tidepoint's headquarters, might include access control, alarm systems, and closed-circuit television. One form of access control is face (or another feature) recognition systems. This allows only authorized personnel to enter, and also serves the purpose of convenience for employees who don't have to carry keys or cards. Cameras can also be placed throughout the building and at all points of entry. Alarm systems also serve as an added measure of protection against theft.

Data storage

Storing data in a secure manner is very important to all businesses, but especially to e-businesses where most of the data is stored in an electronic manner. Data that is confidential should not be stored on the e-business' server, but instead moved to another physical machine to be stored. If possible this machine should not be directly connected to the internet, and should also be stored in a safe location. The information should be stored in an encrypted format.

Any highly sensitive information should not be stored if it is possible. If it does need to be stored, it should be kept on only a few reliable machines to prevent easy access. Extra security measures should be taken to protect this information (such as private keys) if possible. Additionally, information should only be kept for a short period of time, and once it is no longer necessary it should be deleted to prevent it from falling into the wrong hands. Similarly, backups and copies of information should be kept secure with the same security measures as the original information. Once a backup is no longer needed, it should be carefully but thoroughly destroyed.

Data transmission and application development

All sensitive information being transmitted should be encrypted. Businesses can opt to refuse clients who can't accept this level of encryption. Confidential and sensitive information should also never be sent through e-mail. If it must be, then it should also be encrypted.

Transferring and displaying secure information should be kept to a minimum. This can be done by never displaying a full credit card number for example. Only a few of the numbers may be shown, and changes to this information can be done without displaying the full number. It should also be impossible to retrieve this information online.

Source code should also be kept in a secure location. It should not be visible to the public.

Applications and changes should be tested before they are placed online for reliability and compatibility.

System administration

Security on default operating systems should be increased immediately. Patches and software updates should be applied in a timely manner. All system configuration changes should be kept in a log and promptly updated.

System administrators should keep watch for suspicious activity within the business by inspecting log files and researching repeated logon failures.

They can also audit their e-business system and look for any holes in the security measures. It is important to make sure plans for security are in place but also to test the security measures to make sure they actually work. With the use of social engineering, the wrong people can get a hold of confidential information. To protect against this, staff can be made aware of social engineering and trained to properly deal with sensitive information.

E-businesses may use passwords for employee logons, accessing secure information, or by customers. Passwords should be made impossible to guess. They should consist of both letters and numbers, and be at least seven to eight digits long. They should not contain any names, birth dates, etc. Passwords should be changed frequently and should be unique each time. Only the password's user should know the password and it should never be written down or stored anywhere. Users should also be locked out of the system after a certain number of failed logon attempts to prevent guessing of passwords.

Security Solutions

When it comes to security solutions, there are some main goals that are to be met. These goals are data integrity, strong authentication, and privacy.

Access and data integrity

There are several different ways to prevent access to the data that is kept online. One way is to use anti-virus software. This is something that most people use to protect their networks regardless of the data they have. E-businesses should use this because they can then be sure that the information sent and received to their system is clean. A second way to protect the data is to use firewalls and network protection. A firewall is used to restrict access to private networks, as well as public networks that a company may use. The firewall also has the ability to log attempts into the network and provide warnings as it is happening. They are very beneficial to keep third-parties out of the network. Businesses that use Wi-Fi need to consider different forms of protection because these networks are easier for someone to access. They should look into protected access, virtual private networks, or internet protocol security. Another option they have is an intrusion detection system. This system alerts when there are possible intrusions. Some companies set up traps or "hot spots" to attract people and are then able to know when someone is trying to hack into that area.

Encryption

Encryption, which is actually a part of cryptography, involves transforming texts or messages into a code which is unreadable. These messages have to be decrypted in order to be understandable or usable for someone. There is a key that identifies the data to a certain person or company. With public key encryption, there are actually two keys used. One is public and one is private. The public one is used for encryption, and the private for decryption. The level of the actual encryption can be adjusted and should be based on the information. The key can be just a simple slide of letters or a completely random mix-up of letters. This is relatively easy to implement because there is software that a company can purchase. A company needs to be sure that their keys are registered with a certificate authority.

Digital certificates

The point of a digital certificate is to identify the owner of a document. This way the receiver knows that it is an authentic document. Companies can use these certificates in several different ways. They can be used as a replacement for user names and passwords. Each employee can be given these to access the documents that they need from wherever they are. These certificates also use encryption. They are a little more complicated than normal encryption however. They actually used important information within the code. They do this in order to assure authenticity of the documents as well as confidentiality and data integrity which always accompany encryption. Digital certificates are not commonly used because they are confusing for people to implement. There can be complications when using different browsers, which means they need to use multiple certificates. The process is being adjusted so that it is easier to use.

Digital signatures

A final way to secure information online would be to use a digital signature. If a document has a digital signature on it, no one else is able to edit the information without being detected. That way if it is edited, it may be adjusted for reliability after the fact. In order to use a digital signature, one must use a combination of cryptography and a message digest. A message digest is used to give the document a unique value. That value is then encrypted with the sender's private key.

References

Amor, D. 1999. *The e-business (r)evolution*. Upper Saddle River: Prentice Hall.

Beynon-Davies P. 2004. E-Business. Palgrave, Basingstoke.

Chaudhury, Abijit; Jean-Pierre Kuilboer. 2002. *e-Business and e-Commerce Infrastructure*. McGraw-Hill.

Gerstner, L. 2002. *Who says Elephants Can't Dance? Inside IBM's Historic Turnaround.* pg 172.

Paul Timmers, 2000. Electronic Commerce - strategies & models for business-to-business trading, pp.31, John Wiley & Sons, Ltd.

9

E-Government

E-Government is digital interactions between a government and citizens (G2C), government and businesses/Commerce (G2B), government and employees (G2E), and also between government and governments /agencies (G2G). Essentially, the e-Government delivery models can be briefly summed up as:

1. G2C (Government to Citizens)
2. G2B (Government to Businesses)
3. G2E (Government to Employees)
4. G2G (Government to Governments)
5. C2G (Citizens to Governments)

This digital interaction consists of governance, information and communication technology (ICT), business process re-engineering (BPR), and e-citizen at all levels of government (city, state/provence, national, and international).

E-government describes the use of technologies to facilitate the operation of government and the dispersement of government information and services. E-government, short for electronic government, deals heavily with Internet and non-internet applications to aid in governments. E-government includes the use of electronics in government as large-scale as the use of telephones and fax machines, as well as surveillance systems, tracking systems such as RFID tags, and even the use of television and radios to provide government-related information and services to the citizens.

E-Government should enable anyone visiting a city website to communicate and interact with city employees via the Internet with graphical user interfaces (GUI), instant-messaging (IM), audio/video presentations, and in any way more sophisticated than a simple email letter to the address provided at the site" and "the use of technology to enhance the access to and delivery of government services to benefit citizens, business partners and employees". The focus should be on:

- The use of Information and communication technologies, and particularly the Internet, as a tool to achieve better government.
- The use of information and communication technologies in all facets of the operations of a government organization.
- The continuous optimization of service delivery, constituency participation and governance by transforming internal and external relationships through technology, the Internet and new media.

Whilst e-Government has traditionally been understood as being centered around the operations of government, e-Governance is understood to extend the scope by including citizen engagement and participation in governance. As such, following in line with the OECD definition of e-Government, e-Governance can be defined as the use of ICTs as a tool to achieve better governance.

Delivery Models and Activities of e-Government

The primary delivery models of e-Government can be divided into:

- Government-to-Citizen or Government-to-Consumer (G2C)
  - In this model, the G2C model apply the strategy of Customer Relationship Management (CRM) with business concept.
  - By managing their customer (citizen) relationship, the business (government) can provide the needed products and services fulfill the needs from customer (citizen).
- Government-to-Business (G2B)
- Government-to-Government (G2G)
- Government-to-Employees (G2E)

Within each of these interaction domains, four kinds of activities take place:

- pushing information over the Internet, e.g.: regulatory services, general holidays, public hearing schedules, issue briefs, notifications, etc.

- two-way communications between the agency and the citizen, a business, or another government agency. In this model, users can engage in dialogue with agencies and post problems, comments, or requests to the agency.
- conducting transactions, e.g.: lodging tax returns, applying for services and grants.
- governance, e.g.: To enable the citizen transition from passive information access to active citizen participation by:
  - Informing the citizen
  - Representing the citizen
  - Encouraging the citizen to vote
  - Consulting the citizen
  - Involving the citizen
  - Non-internet e-Government

While e-government is often thought of as "online government" or "Internet-based government," many non-Internet "electronic government" technologies can be used in this context. Some non-Internet forms include telephone, fax, PDA, SMS text messaging, MMS, wireless networks and services, Bluetooth, CCTV, tracking systems, RFID, biometric identification, road traffic management and regulatory enforcement, identity cards, smart cards and other Near Field Communication applications; polling station technology (where non-online e-voting is being considered), TV and radio-based delivery of government services (e.g., CSMW), email, online community facilities, newsgroups and electronic mailing lists, online chat, and instant messaging technologies.

GLOBAL E-GOVERNMENT LANDSCAPE

Since the mid-1990s governments around the world have been executing major initiatives in order to tap the vast potential of the internet for the distinct purpose of improving and perfecting the governing process. Like the personal computer, the internet has become an indispensable tool in the day-to-day administration of government. In an effort to gain an appreciation of the global e-government landscape in 2001, the American Society for Public Administration (ASPA) and the United Nations Division for Public Economics and Public Administration (UNDPEPA) undertook a research study analyzing the approach, progress and commitment on the part of the 190 UN Member States. The factors of the study report —'Benchmarking E-government: Assessing the Progress of United Nations Member States'— is elaboratively discussed in this section.

National e-government program development among the UN Member States advanced dramatically in 2001. Countries whose web presence in previous years consisted of one or two static government web pages began offering content rich, well-designed, citizen-centric sites. But despite creative initiatives, national e-government program development remains overwhelmingly at the information provision stage. The level of sophistication in which countries are using the internet to deliver quality information does, however, vary considerably.

Table 1: Online Profile of UN Member States

UN Member States:	190
with a government website presence	169
with a National Government Website	84
with single entry portals	36
with sub-national govt websites	84
with online transaction capacity	17

Full-fledged commitment to e-government implies that a country's leadership recognizes the fact information has become a social and economic asset just as important and valuable as traditional commodities and natural resources. Information benefits the most the individuals and industries which have unimpeded access to its acquisition, and the self-determination to convert essential data into knowledge.

In 2001, of the 190 UN Member States, 169 (88.9%), of their national governments used the internet in some capacity to deliver information and services. For 16.8% of these governments, their presence on the internet was just emerging. The official information offered in these countries was often static in content and limited to only a few independent websites.

A country's social, political and economic composition most definitely correlates closely with its e-government program development. However there were exceptions, as evidenced by several developing and transitioning economies. Key factors such as the state of a country's telecommunications infrastructure, the strength of its human capital, the political will and commitment of the national leadership and shifting policy and administrative priorities play important roles. Each of these factors influence how decision makers, policy planners and public sector managers elect to approach, develop and implement e-government programs.

E-government can offer numerous possibilities for improving how a nation's public sector responds to the basic needs of its citizens. There is however a wide variance as to the process in which governments choose to realize such potential.

National governments play four distinct roles in an information society:

— Determine the policies and regulatory structures

— Deliver the programs and services of government to the citizen

— Use the information infrastructures to enhance the internal administrative practices

— Interface with citizens in the democratic process of government.

Over the past ten years the dramatic advances made in information and communications technology (ICT) have transformed much of the world into a digitally interconnected community that is increasingly functioning on a in 365 / 24 / 713 basis. Throughout this period, especially the past five years, the predominant drivers of change have been the internet and the world wide web. Both have added a new, and arguably indispensable "e" dimension to academia, commerce, and now government.

Considerable resources, both human and financial, are being committed to launching and perfecting programs which deliver government information and services online. Such initiatives are intended to improve administrative operations and enhance government efficiency while realizing the residual effect of deeper citizen involvement in the governing process. From Armenia to Zambia the concept of e-government is being openly embraced.

For most of the UN Member States, electronic service delivery, or e-government, is still a new and challenging medium. To be precise, e-government encompasses many applications and incorporates virtually all ICT platforms. However, it is the internet that is the most widely recognized and identifiable component driving e-government. A major indicator of a societies' openness is access to the information available on the internet. Restrictive states have sought ways to contain or manage information and access, ultimately compromising an enabling e-government environment.

The year 2001 saw a greater expansion in government online presence than the previous five years combined. Website content and online applications progressed from static, public affairs life-brochures to virtual information centers where the interaction between citizen users and the public sector is continuous. Last year also saw an increase in the online transaction of business with government. In 2001, it was estimated that globally there were well over 50,000 official government websites with more coming online daily. In 1996 less than 50 official government homepages could be found on the world-wide-web.

Of the 190 UN Member States, 169 were providing some degree of information and services online. Based on the available resources, policy priorities and political commitment, however, the scope and content of national e-government programs contrast substantially. Namibia's Ministry of Health and Social Services for example, maintains an effective information gateway despite extremely inadequate resources, while the Republic of Koreas National Tax Service is indicative of what can be done with unabridged commitment and support.

E-government development is constant and conspicuous. It has received considerable attention through a steady stream of events at the national and international levels. In March 2001, for example, the Third Global Forum on Reinventing Government chose for its theme: 'Fostering Development through E-government'. Participants from 122 nations, which far exceeded the combined official representation of the first two Global Forums, met in Naples, Italy to share best practices, information and ideas.

The Third Global Forum articulated the following key points:

— E-government can consistently improve the quality of life for citizens and can create a sharp reduction of costs and time.

— E-government will eventually transform the processes and structures of government to create a public administration less hierarchical, empowering civil servants to serve citizens better and to be more responsive to their needs.

— E-government must be given serious consideration also in the developing countries not only for its potential for stronger institutional capacity building, for better service delivery to citizens and business (thus increasing local social and economic development), for reducing corruption by increasing transparency and social control, but also for 'showing the way' to the civil society and business community.

Wide-ranging e-government programs remain at, or near, the top of most countries' policy agendas. For many nations, digitizing service deliveries like filing personal income taxes online or paying VATs electronically represents a marked departure from the traditional paper-based way of doing business. For some countries, such departures have culminated in success. For others the challenge is formidable, but not insurmountable.

The basic principles of e-government are:

—Build services around citizens' choices

—Make government and its services more accessible

—Facilitate social inclusion

—Provide information responsibly

—Use government resources effectively and efficiently.

E-government potentially empowers individual citizens by providing them with an alternative channel for accessing information and services and interacting with government. It also gives the individual citizen another choice: whether to become an active participate in the governing process or remain a passive observer. Providing citizens with new choices is a goal that resonates throughout many of the national e-government strategic plans.

Both the intended and the residual outcomes of this objective are considerable: open communication, enhanced transparency, increased social inclusion and citizen participation, democratic enrichment and superior governance. These outcomes however, tend to be more potential then extant, and objectively quantifying a potentiality is in itself a test.

Public Face of E-Government

Virtually all government services can be classified under one of three fundamental categories: informational, interactive and transactional. The first, informational, is by far the most significant. Information is at the heart of every policy decision, response, activity, initiative, interaction and transaction between government and citizens, government and businesses and among governments themselves. How information is collected, processed, analyzed, packaged and disseminated is in itself a specialized industry.

Successful citizen-centric e-government programs emphasize the indispensable nature of information while balancing its often limited shelf-life and considerable inflationary component. In the information and knowledge age, there is no institution that produces raw data and new information with more regularity than government.

Since services are the public face of government, the primary objective of all e-government initiatives is to provide the citizen user with an efficient alternative medium for interacting with public sector service providers. This is generally accomplished by improving the flow of information both externally and internally.

Eighty-eight percent of the UN Member States have made at a legitimate effort to commit to some form of e-government; that is 169 countries have an established online presence with official government websites. However, in 2001, for over 25% of the countries, the content of official websites consisted of static and insufficient information often of a public relations nature and consistently with strong political overtones. Such sites can hardly be described as service delivery or considered citizen-centric since they are not a medium to elicit useful feedback.

This trend is predominant in a several emerging countries in Asia, the Caribbean and throughout Sub-Sahara Africa where countries like Zimbabwe exploit the web for very singular purposes. The industrialized countries have taken a less desultory approach in developing their programs and the types of services they are providing. This is particularly evident throughout Europe, North and South America as nearly all the countries in both regions have highly dynamic and interactive official government websites, where content is accurate, specialized and regularly updated.

The capacity to conduct transactions online at the national government level in 2001 was available in 17 countries. There was also considerable activity at the sub-national level, which would suggest that in countries where e-government is evolving autonomously or without a coordinated national strategic program, transactional service delivery will occur irregularly with, in many cases, local governments demonstrating such a capacity before the national government. Although online transactions are one of the primary features that justify a wide-ranging e-government initiative, it is worth noting, that despite the benefits of technological and the impact of globalization, the national economies of over 75% of the countries indexed are substantially cashed based.

In the majority of these societies, credit-card use is still reserved for a select and privileged minority. Consequently, the necessity for online transactions in such countries may not be as great as the need for reliable information. Credible information is sine qua non for maintaining a balanced and open dialogue between decision-makers and the civil society. The policy participation process is one example.

A higher level of participation that goes beyond just providing feedback and comment is the participation of citizens in the processes of policy development and decision making of government. Traditionally this has been practised through the use of voting and referenda and again mostly at community and local levels. It goes beyond simply providing feedback; rather it is a process of discussions and negotiations which often involves personal interaction.

E-government potentially increases citizen involvement in the process of governance at all levels by introducing new voices to the dialogue through online discussion groups, thus expanding outreach and influence while enhancing the rapid development and effectiveness of interest groups.

In the United States for example, there is concern among academics, activists, and elected officials that government websites might focus more on providing services, and less on facilitating civic involvement. This type of service orientation, they argue, treats citizens as consumers rather than partners in government, and thus inhibits public engagement with the nation's political environment.

While users are certainly taking advantage of all the services and information made available on government sites, a smaller portion are active in using the Internet to monitor public affairs. In 2001, an estimated 42 million Americans (24% of those with access) used government Web sites to research public policy issues; 23 million (13% with access) used the Internet to send comments to public officials about policy choices; and 13 million (7.5% with access) participated in online lobbying campaigns. Despite the incredible advances made in information technology, digitizing government is, and will continue to be into the foreseeable future, a complex and constant process. Countries vary radically in their approach, level of development and overall commitment to e-government.

Critical endemic factors like available resources, political leadership, economic capacity and the character of the civil society deeply impact on the scope and breadth of a government's e-gov policy. How a nation ultimately shapes its e-government commitment ideally should consider these factors and respect the citizen-centric approach rather than being influenced by short-lived trends or what outcomes may be politically expedient.For countries, like the United Kingdom launching an e-government program is an extensive, meticulously planned exercise with ambitious goals and targets. The UK also enjoys the advantages of unabridged political support. For other nations, especially those in Sub-Sahara Africa, just going online can be a frenetic leap from the past into the future with little time to absorb the present.

E-government is a new and for the most part a nascent activity. Program success is not necessarily measured by the increase in the number of government websites, the number of daily 'hits' or user visits to each site, or by the number of officials who can be accessed by e-mail.

Successes are kept in perspective, and setbacks are to be expected. Despite the highly visible and multi-functional sites, many governments continue to "flight test" their programs in an attempt to find the right combination of services, features, content and entry points that are efficient, cost-effective and truly citizen-centric. However, time in cyberspace does not allow governments to celebrate very long their accomplishments, nor mull over their setbacks.

Stages of E-government Development

Benchmarking is an essential and valuable exercise since the practice allows individuals to mark a specific policy or program's progress and chronicle successes and setbacks over a given period of time. This provides policy developers, decision makers and public sector managers with vital information in order to make mid-course adjustments, improve program efficiency and ensure that the tax-payers are getting a return on their investment. Several countries, international organizations and private sector consulting firms have developed useful tools in an attempt to measure e-government progress through a series of targets that mark a specific stage of development.

Using the world wide web, all 190 UN Member States were accessed and over 1900 national government websites examined in order to analyze and chart each country's level of progress. The citizen-centric approach was the basis for the analysis. That is sites and sectors considered to be most representative of services that the majority of citizens were likely to seek were evaluated. The primary or target sectors were: health, education, labor or employment; social welfare and services, and finance. Countries were also evaluated for a capacity to conduct transactions online. Countries were assessed by the following criteria:

— An official government web presence must exist.

— The type of service delivery available: basic or informational, interactive, transactional.

— The presence and of services in five critical sectors: education, health, labor/employment, welfare/social services and financial services.

— Use of single entry portals; and de facto portals.

— To a lesser degree, fidelity to strategic plans, use of e-gov teams.

National government strategic planning, characterizes e-government development as a linear progression, with service providers moving through four, five or even six levels or stages before achieving the stated program objectives. This type of benchmarking is a based primarily on analyzing website content; any interactive features (e-mail), quality and timeliness of information and the capacity to conduct online transactions. It provides a convenient reference point for developed and emerging nations, but it presupposes a definitive level of technical sophistication.

E-government programs are subjected to a number of internal and external factors. Technology and trained staff obviously play a vital role in e-gov development and influence how a government will implement its program. But it is the availability of resources that dictate the goals of a particular government. Ramping up from level one to level two, level three, etc is, for now, a consequence of the patterns of contemporary program evolution. This does allow policy planners a degree of flexibility and creativity in program development while still following the game plan. Based on the availability of a governments' resources, an effective program can be designed and launched with the intent of being fully transactional before ever going online. Ramping up also allows the luxury of utilizing a cost-effective practice of testing through pilot programs.

The website research was conducted during a three month period from May to July and repeated from October to December, 2001 in order to measure progress and ensure accuracy. Each Member State was examined using sites available on the world wide web during these periods. As stated earlier, e-government development can be swift and continuous, yet inchoate. Change and improvement must be a permanent part of the process if a country is to achieve the stated goals within its strategic framework and to offer the most inclusive citizen-centric approach.

Progressive governments are upgrading their sites regularly. Expanding the types and quality of their online services and improving their content daily in an attempt to achieve the highest measure of user satisfaction, administrative efficiency and cost effectiveness. The website analysis conducted as part of this research is an appraisal taken at a particular period in time --- a digital photo of the global e-gov landscape during 2001. For the purpose of the UN/ASPA survey, a country's online presence can be categorized by one of five stages: emerging; enhanced, interactive; transactional; and fully integrated or seamless.

Much of the current performance analysis recognizes e-government development as a linear progression, with countries moving through four, five or even six levels or stages of advancement. This form of benchmarking is a based primarily on analyzing website content, special features, the quality and type of information offered and the capacity to conduct online transactions. It provides a convenient reference point for developed and emerging nations, but it presupposes a certain level of sophistication.

For most countries, gradually ramping up to a more advanced level or stage of development is a sound strategic approach. Factors like technological improvements, financial and human resources, political commitment and citizen participation, all contribute to how quickly an e-government program progresses.

There are, however, other factors that should be noted as well. In certain developing and emerging nations current cultural and economic conditions may justify e-gov programs that reflect the immediate needs and technical capacity of these societies. In such cases, attaining level three or even level two may be a realistic strategic development goal and a sufficient online response.

Consequently, some programs may seem modest in comparison to progressive or industrial nations. It should be emphasized that the primary objective of any e-government program is to address the needs of citizens. For a number of countries attaining the enhanced or interactive level may successfully fulfill this objective. There are a finite number of services that citizens and businesses can transact online with national governments. Transactional services vary depending upon the type of political system.

Highly centralized governments will retain jurisdiction over many transactions that decentralized and federal systems will devolve to sub-national levels.

Table 2: Stages of E-Government

Emerging:	An official government online presence is established.
Enhanced:	Government sites increase; information becomes more dynamic.
Interactive:	Users can download forms, e-mail officials and interact through the web. Transactional: Users can actually pay for services and other transactions online.
Seamless:	Full integration of e-services across administrative boundaries.

Emerging Presence

Nearly all 32 countries, (16.8 of those surveyed) at the Emerging Presence Level, are among the world's least developed nations with over half (18) in Sub-Sahara Africa. Countries at this level averaged 3.9 official websites per government (in 2001, the United States had over 31,000 Federal, State and local sites) with content consisting of predominately static and insufficient information that is updated infrequently, if at all. Most programs are in their embryonic stage, providing few interactive features like e-mail or downloading forms.

Actual online services, as of this writing, are non-existent for the countries at this level, and no country offered an official website for any of the five target sectors. Official information for many of these countries is predominately of a highly partisan, political nature, i.e. bios and predictable spin on the prime minister or the party in power. The number of sites disproportionately favor the political parties over the service-provision or administrative sites.

There is little or no information that would assist an individual in accessing a specific public service. For the most part, content could hardly be described as citizen-centric. Sectors with sites that are funded by international organizations or outside groups, or those promoting foreign investment and tourism tend to be more highly developed as a result of the political and economic weight placed on such issues. Some island nations, whose primary source of economic activity is tourism, have one or two sites promoting the advantages of visiting or investing in their country. Seychelles and Trinidad and Tobago are such examples.

Obviously many of these countries are among the poorest in the world, whose limited resources are stretched well beyond their finite capacities. National priorities focus on basic survival issues. However some countries have displayed an intrinsic capacity for progress. Gabon's national government site is an example of a de facto single entry portal and allows users to access official national government sites. Guiana and Botswana are on the verge of ramping up to the next level.

Most government sites do not have links to all of the ministries and agencies, and many have not been updated. The sites typically are not used for administrative purposes. For the countries at the Emerging Presence level, e-government program development is erratic and disparate. There is little evidence of any coordination or strategic planning within ministries or departments, let alone, national governments. However, this condition may be reflective more of the prioritization of needs rather than organizational or political indifference.

Enhanced Presence

Of the 65 countries (34.2%of the Member States) with an Enhanced Presence, 58% are emerging or newly industrialized economies, and, with the exception of North America, are found in all geographic regions. The average number of sites for countries at this level is 18.

There are several cases of developing countries like Cote'd Ivoire, Burkina Faso, Ghana who are quickly succeeding in their e-government programs as evidenced by their commitment to content improvement. There is still, however, very little if any online representation of ministries or departments dealing and with social welfare, education or health issues.

The potential for more advanced development in e-government is apparent from some areas of Ghana's site, however. Administrative guidelines and regulations for obtaining a license to export cocoa are clearly outlined in extensive detail. Although the sites do not allow the user to download the necessary forms and applications, the information is easily accessible and thorough. The majority of CIS states including Georgia, Kazakhstan, Turkmenistan are at the Enhanced Presence level.

Presently, Nigeria's government web sites are primarily for public affairs issues, with very little dynamic information to the citizens of Africa's most populous nation. Political agendas of the party in power are obvious and appear on the forefront of the site. In Nigeria's case, the unofficial government web site is the Presidency.

Quotes about combating corruption and allegiance to the presidency appear on the page. The site itself does not provide direct access to all ministries and legislative or judicial issues, including laws and regulations or court decisions. This is not uncommon among the countries at this level, particularly those in Africa.

Nigeria has no ministries online, and instead provides links to an array of agencies and departments concerning export regulations, ports authority, public enterprises, corporate affairs, and investment promotions. The focus is clearly on promoting investment and enhancing economic growth and development.

However, all sites or information relating to education, social services, health care or women and children specifically are conspicuously absent. This pattern of unbalanced implementation is common at this level and among developing countries.

Several Central American countries, such as El Salvador, the Dominican Republic and Guatemala, despite persistent program obstacles (i.e. technological, financial and human resources), are on the verge of ramping up to the next level of development. The Principalities of Monaco and Liechtenstein each have technically advanced websites, but their content provides general information of a public affairs nature promoting the benefits of their countries.

The majority of nations at the Enhanced Level show a greater degree of diversity in their online content and the quality of information. Although the content on most official sites is predominantly political or public affairs in nature, more than half the countries are developing a citizen-centric approach with sites for at least one of the targeted sectors. Forty-two countries are upgrading their information on a regular basis.

Interactive Presence

The Interactive Presence clearly embodies the citizen-centric approach as content, information and services become what the people should expect rather then what governments prefer to offer.

The delivery of information and services is aimed at maximizing the importance of the consumer by ensuring that ease of use is priority. Portals are the preferred point of entry; content is critically managed and information is well balanced. Security and privacy features are prominent among sites.

At this stage, perhaps the most varied profile of the Member States in terms of national economic, political and social conditions can be found. Of the 55 countries who offer interactive services, 20% are developing nations. The remaining 80% are evenly divided between newly emerging economies and industrialized countries. Virtually all the governments have demonstrated a strong national commitment to a citizen-centric e-government program and are well advanced technically in their programs. Only ten countries did not have websites for all five target sectors. Each of the ten, did however have a minimum of three target sectors online.

Twelve countries (24%) at this stage employ official single entry portals. Thirty-one countries (62%) have official national government sites which serve as de facto portals, each linking the user with most ministries, agencies departments as well as government and elected officials. Of the 31 countries with official national government sites, 24 link to the ministries of the five targeted sectors: health, education, labor, social welfare and finance. Communicating with government or elected officials is available through e-mail and post comments features for each of the 55 countries.

Information and content take on a greater significance at this level and are continuously updated. The sites are also upgraded on a regular basis. Although there is always a degree of political information present in the content, it does not dominate a country's official government web presence.

In all 55 countries, users have the capacity to download and request either documents or forms from a specific ministry, department, agency or unit. For a number of countries like the Netherlands, Sweden, Japan, ramping up to the transactional level is only a matter of time,

As a gateway or single point of entry to government services, portals are becoming the standard. In 2001, 36 countries provided easy access through single entry portals. At the regional or state level the number is much higher. All 26 of Brasil's states use this approach, as do Germany's Landers, the majority of Mexico's states and all 50 of the US states. Serving as much more than a simple gateway, however, a portal offers an opportunity to reorient services around the needs of citizens while consolidating back office responsibilities. The United States and the United Kingdom are excellent examples. Eighty-four countries have an official national government web page. These sites can serve as de facto single entry portals.

Although online transactions are one of the primary services that demonstrate e-government's utility, the point must be reiterated, that despite daily advances in information technologies and the inescapable impact of globalization the national economies for the large majority of UN Member States are predominately cashed based.

Credit-cards, online banking and other paperless forms of e-commerce are, for the present, available to those who have the means to access such services. The necessity for online transactions in many of these societies may not be as great as the need for reliable and practical information.

Transactional Presence

At the Transactional Presence level, a country has fully demonstrated the capacity for users to interact with the government by purchasing publications or other item, or utilizing a service like obtaining a passport and paying for it online with either a credit card, bank debit or by some other electronic means. Australia, the UK, Ireland, Brazil and the United States all present a concise, easily navigable portal to its transactional services. Perhaps the most frequently cited online national government transaction is the filing and payment of taxes.

Online taxation payment is technically in the pilot program stage for nearly all 17 countries at the transactional level. Spain, has one of the most sophisticated online taxation programs in Europe. Citizens can pay income taxes online, and businesses can pay income, property and sales taxes online as well. Spain also features the use of digital signatures.

Germany and Finland have successfully launched payment programs for citizens as well as businesses. The Republic of Ireland has introduced online VAT payments. Ireland also provides services like paying one's utility bill, which is available at most General Post Offices throughout the Republic, now online. Norway is the site of the Norwegian Revenue Service offers several transactional services, including the payment of income, property and sales taxes.

The United States improved its online taxation payment program from a year ago, though the number of those filing e-returns has remained below expectations. France has used a variation of electronic tax filing for several years. However the transition to online taxation has been slow to gain acceptance as those wishing to use the service must negotiate several administrative steps. Mexico has instituted a similar system.

Canada has perhaps one of the most comprehensive e-government programs. In addition to offering a bi-lingual portal, it provides citizens with a wide range of services in a highly efficient and user-friendly manner. Content and services reflect the government's unremitting commitment to improvement and to providing the best product to Canadian citizens and international users. Level Five or Seamless government represents, for some countries, the ultimate goal. However very few countries have publicly acknowledged it as a final policy objective.

The United Kingdom and Singapore are two examples. Seamless government is not as easy to qualify as the four previous levels. It is characterized by the objective of having all online services, information, websites, etc available to the user through a single entry point that is driven by a super-search engine. But that's only half the equation. It also presupposes a reorganization of internal administrative structure of government's responsibilities to process these services: in effect eliminating administrative boundaries.

It is sound strategic thinking, however it is a perfect world scenario. A more futuristic vision of service delivery based on the presumption that both technology and human nature will be so compliant that the administrative procedures will transverse organizational boundaries and become one simple seamless process. Attaining this level would necessitate a considerable degree of political, administrative and managerial cooperation. Before the government's time and the public's resources a specific set of environmental conditions must exist. These include:

— A realistic political vision and plan that completely grasps the strengths and weaknesses of its public sector's capacity.

— A fully committed national leadership secure enough to sustain the political opposition to such a reform program.

— A confident and professional administrative culture willing to relinquish some degree of organizational and administrative territory.

For the majority of countries, attaining the goal of seamless government may be an abstract or remote objective.

Website Evaluation

Governments analyze and test a variety of measures and indicators in order to find the most practical benchmarking methods. Website Evaluation should be used in conjunction with other assessment and benchmarking activities that compliment its value. When perfected, website evaluation can be useful in fulfilling such governance goals as:

- Improved planning and goal setting of E-government initiatives.
- Improved decision-making and resource allocation related to E-government programs.
- Determining the effectiveness of a E-government website.
- Determining the degree to which the website adds value to the organization.
- Providing trend data to assess change over time.
- Contributing to continuous improvement efforts and benchmarking.
- Identifying problems and possible solutions, as well as the effectiveness of corrective action.
- Empowering organizational actors to seek and enact solutions.
- Developing accountability organizational learning.. Improving public information access services of those websites.

The E-Government Index

To ensure that a national e-government program realizes its maximum potential, the existence of a favorable or enabling environment is paramount. By regularly assessing the core areas that are requisite for sustaining an information society — institutional capacity, cultural and human resources conditions, ICT strengths and the political commitment — governments would be in a position to carefully evaluate performance opportunities and challenges while candidly appraising their strengths and weaknesses. E-capacity analysis allows policy and decision-makers to initiate the appropriate actions necessary in order to ensure the needs of their citizens are effectively met. National governments play four distinct roles in sustaining the enabling environment of an information society:

- Determine the policies and regulatory structures.
- Deliver the programs and services of government to the citizen.
- Use the information infrastructures to enhance the internal administrative practices.
- Interface with citizens in the democratic process of government.

In order to effectively respond to the numerous technological, fiscal and cultural factors that contribute to developing and sustaining an information society, decision-makers and public sector professionals should maintain a clear, practical vision of their national e-gov programs that realistically reflect the strengths, weaknesses, needs and priorities of their constituents. The are several core factors that have been identified in UN reports on ICT and have also been discussed in various forums impacting developing countries in ICT and e-government projects.

Table 3. ICT and e-government projects: Strengths, weaknesses, needs and priorities

Core Factors	*Symptoms*	*Consequences*
Institutional Weakness	Insufficient planning systems	Inadequately designed
Human Resources	Unclear objectives Shortage of qualified personnel Lack or professional training	Cost over-runs Insufficient Support Isolation from sources of technology
Funding Arrangements	Underestimated project costs Lack of recurring expenditure	Unfinished projects Higher maintenance costs
Local Environment	Lack of vendor representation Lack of back-up systems	Lack of qualified technical support Implementation problems
Technology and Information Changes	Limited Hardware / Software Inappropriate software	Incompatibility system Over-reliance on Customer applications

If a national government's strategic 'vision' anticipates greater citizen participation, improved service delivery, increased efficiency, administrative modernization, enhanced transparency and increased foreign investment, then the political leadership must be fully prepared to see such a commitment through to the completion of these objectives regardless of the challenges.

The strategic goals of any national policy should, most importantly, reflect the needs and expectations of the citizens, not necessarily the grandiose dreams of policy advisors or ephemeral promises of the political leadership. Like resources and technical capacity, priorities and expectations vary widely from county to country, government to government and citizen to citizen. Where perfecting the payment of taxes online is one of several principle objectives in Spain, for example, simply 'going online', or staying online may be the most immediate goal in Senegal.

Table 4: Global Online Population

Online	*August ,00*	*August ,01*	*% change +/-*
World Total	408 mil	514 mil	+ 20
Africa	3.2 mil	4.2 mil	+ 24
Asia /Pacific	105 mil	143 mil	+ 28
Europe	114 mil	154 mil	+ 26
Middle East	2.5 mil	4.7 mil	+ 47
Canada /USA	168 mil	181 mil	+ 8
Latin America	17 mil	25 mil	+ 32

In developing and transitioning countries, factors such as institutional weakness, lack of qualified or properly trained staff and technological limitations potentially make even a minor e-gov activity, like responding to e-mail, a major administrative challenge.

The need for technical assistance, administrative capacity building and human resource training are the most conspicuous areas where the international community and the United Nations can play an active, facilitating and significant role.

The web, being a transparent and highly visible medium, is the public's most palpable assessment of a country's e-readiness. Consequently a subtle dynamic is taking place among governments: the allure of the status associated with the recognition (or the perception) of being e-ready.

In order to realistically compete (or at least be given the opportunity to compete) as a knowledge-based economy in the global e-commerce arena, many developing country governments are committing to fast track applications that are either lack real substance or address only the needs of a single sector such as the business community, often at the expense of a citizen-centric approach.

Compiling the E-government Index

There are several key factors that define the core areas of an enabling e-government environment. Many are endemic to a nation's individual political, economic and cultural fabric. In short a country's national character. Yet because the visions, goals and policies that encompass e-government vary considerably among practitioners and users, comparative indicators may not always be precise, while any proven reference points, at this time, are virtually nonexistent. Consider also the fact that despite a pattern of continuous growth in internet usage it is too early to anticipate with confidence what percentage of a country's population will completely accept e-government and make it a regular part of their lives.

Several proven indicators, however, do exist that are representative of a country's capacity to launch, sustain, perfect and promote an effective e-government program --- a country's e-gov environment--- and can be used as dependable benchmarks. Telecommunications indicators like, pc's, internet hosts, telephone lines, are indispensable, as without these requisite technologies, a country's capacity to sustain online service delivery is nonexistent. Analyzing website content, access patterns, online services and official information, give only a partial (albeit a critical) account of a countries overall e-gov environment.

In devising a methodology for benchmarking and performance measurement for an enterprise as nascent and unique as e-government, one approach would be to include many variables in order to obtain the most comprehensive picture as possible. But currently, there are only a few comparable universal indicators suitable for quantifying e-government.

Such comprehensiveness may not be entirely desirable. While too few indicators may tell only a partial story, too many can distort the picture or dilute general trends. The model practice will weigh disparities and balance inequities.

The E-government Index attempts to emphasize and balance the conditions that are most representative of a country's capacity to develop, sustain and provide unimpeded access to timely, useful and relevant information and services for every segment of its population. In order to best capture a balanced interpretation, three distinctive measures consisting of primary indicators available for most of the UN Member States were developed. The measures are:

— Web presence Measure;

— Telecommunications infrastructure Measure and;

— Human Capital Measure.

The first —web presence measure—captures a country's online stage of development have discussed earlier. The second measure—telecommunications infrastructure measure —compares six primary indicators which define a country's ICT infrastructure capacity. The sources for the statistics are the 2001 International Telecommunications Union Report and the 2001 UNDP Human Development Report. They are:

— *PCs per 100 individuals*: For now, PC's are the primary device for accessing the internet until access becomes universally available through other mediums like television; this statistic is fundamental in quantifying a country's capacity to deliver online service.

— *Internet hosts per 10,000 individs*: Measures internet penetration. Obviously, the greater the number of internet hosts and service providers, the greater the opportunity for citizen access.

— *Percentage of a nation's population online*: Estimates how many are citizens are using the web. The source for these figures is the only organization that provides statistics for nearly every UN Member State

— *Telephone lines per 100 Individs*: Which is basic infrastructure measure. The greater the number of telephone lines the likelihood increases for access.

— *Mobile phones per 100 individs*: Indicates a country's potential for wireless capacity. Wireless connectivity is extremely important in many developing countries. This could play an important role in the near future, as mobile access becomes more comprehensive.

— *Televisions per 1000*: This indicator was included to assess the prospects of web TV. Cable and satellite TV potentially offer the highest rate of access of any hardware device.

Having the education, freedom and desire to access information is critical to e-government's efficacy. Presumably, the higher the human development the more likely citizens will be inclined to accept and use e-government services.

Having the technical means to access relevant and timely information and services is a critical factor addressed in the web presence and infrastructure measures. The third measure— human capital— attempts to capture a country's, and its citizens' facility, opportunity and willingness to use online government. The three indices are:

- *The UNDP Human Development index,* which measures a society's well being, including level of education, economic viability and healthcare.
- *The Information Access Index* which draws on two annual surveys from Transparency international and Freedomhouse International.
- *Urban / rural population ratio* which gives an indication of internet service patterns and how access may be prioritized. Countries with a greater percentage of its inhabitants in rural areas may find bringing service to its population a greater challenge.

Global overview

Overall, 61 or 42% of those countries indexed placed above the global mean of 1.62. Thirty-six countries, or 25% of the 144 indexed have High E-government Capacity, exceeding the value of 2.00. The majority of the countries with a High E-gov Capacity (25) are OECD members. Countries indexing above 2.00, predominantly, have the technical and human capacity and the resources to sustain innovative and productive citizen-centric e-gov programs with few encumbrances.

The United States (3.11) is the overall global leader and the only country to register an index greater than 3.00. Clearly, the US's tradition as a pioneering leader in the field of information technology, along with its well-educated population, economic strength and a fidelity to the principles of freedom of information give it a significant competitive advantage.

But despite an abundance of technological and human capital reserves, the US was slow off the e-gov mark and only began to make serious strides in 2001 with the performance of Firstgov.gov, the official portal of the federal government; improved online transaction capabilities and the US Government's remarkable ability to disseminate accurate, reliable and timely information immediately following the events of Sept. 11th, 2001 and thereafter.

Several countries, such as Mexico (2.16), Brazil (2.24) and Chile (2.03) for example, were able to overcome persistent infrastructure limitations, like inadequate hardware availability or scarce internet access, and rigorous human capital challenges, to develop complete e-government programs. In Mexico and Brasil, much like in Australia, the United Kingdom, Ireland and Sweden the commitment on the part of political leadership has been strong and constant.

Twenty-six countries (18% of those indexed) have Medium E-government Capacity and placed between 1.60 and 1.99 with Poland (1.96) just missing inclusion among the global leaders. Countries in this group are predominantly technologically advanced and in many cases enjoy a well educated population. However, each country may be weak in one or more key individual areas. Poland, for example, scored high in its web presence measure suggesting an unabridged commitment on the part of the political leadership to a competitive e-government program.

However in the infrastructure and human capital measures, Poland ranked comparatively lower than most of its European neighbors. Egypt (1.73) is another example where the commitment to an effective and efficient e-gov program is a priority despite ongoing infrastructure and human capital limitations.

A greater number of countries, 36 (25%), demonstrated a Minimal E-government Capacity indexing between 1.00 - 1.59 than any other level. Included in this group were newly emerging economies, developing and least developing countries. The e-government programs among these countries reflect the realities of their limited infrastructure and human capital capacities while balancing the priorities of their development needs. Information provision is the primary service provided by governments and in several cases it is of minimal value to the users.

Lack of basic assets like PC's, partial internet access, and an inadequate number of telephone lines are part of the equation. Limited economic resources are the other part. Despite inadequacies in the areas of education, human capital levels are improving.

The commitment to a fully functional e-government program, however, is strong in many countries including, Armenia (1.59), Costa Rica (1.42), Jamaica (1.31) and India (1.29).

Among the developing countries, India, has benefited from the combined effort of its political leadership and highly professional civil service in developing a multi-functional e-gov program. It maintains a single entry portal and offers many of the requisite features found in the programs of the global leaders. Of the 35 (24.3%) countries who indexed below 1.00 --- Deficient E-gov Capacity --- 31 are among the world's least developed nations with 25 from Sub-Saharan Africa. For the majority of these countries, e-government registers a low order of priority on the policy agenda.

Augmenting ICT capacity must be rationally balanced with food, medical, educational and labor issues. However, there are nations who are finding the resources and the expertise to make e-government happen.

Bangladesh, Kenya, and Ghana, for example recognize the importance of the role technology plays in development and have embarked on e-government programs representative of their respective development agendas.

Public E-administration

Information age governing presents an entirely new set of challenges for decision-makers, public sector professionals and citizens. How individuals and businesses interact with government is being fundamentally altered by the technological advances driving e-government. This transformation has public sector professionals considering questions like:

— How will e-government affect the performance of public organizations?

— What are the structural effects of e-government and information technology on the public organization?

— What skills do public employees need in order to maximize their performance in an information age?

— What new leadership skills will be needed in the e-governing age?

— Will e-government instill individuals in public sector organizations with a greater degree of autonomy, enabling them to re-think conventional administrative practices?

Developing an effective online public administration or e-administration means balancing the needs of two constituent groups: one external --- the citizens or the customers, and one internal --- staff and management, or the administrative back office. These two obligations are neither conflicting nor mutually exclusive. In both situations the internet has become essential in augmenting the administrative system in support of its mission.

For a number of countries (industrialized, emerging and developing) there exists a propensity to center their e-government projects and budgetary resources on the output of services provided to external users before ensuring the administrative capacity exists to support such initiatives. The reason for this choice, as a recent EU report found, is, is the need to catch-up that prompts governments to go as fast as possible giving priority to matters of direct interest to citizens, before being fully capable of providing such services. The obvious result is that there is limited back office capability to handle the new responsibilities created by e-government, thus potentially compromising online service delivery efficiency.

In developing countries the chronic lack of qualified staff and inadequate human resources training has been a problem for years. The new e-government programs that many developing countries feel compelled to launch further compounds this problem. Countless official guidelines and methodologies have been published to help countries implement e-government initiatives more successfully. While these tools have been used with some success, many e-gov programs are underachieving and falling short of their initial promise. Implementation problems exist because too many organizations conceive of, organize, and implement e-government programs first and foremost as generic information technology projects.

The most perplexing problems, however, are almost always the ones created as a result of the politics of organizational change. If change-related issues get the kind of consideration they warrant, then implementing e-government programs will be a much less complicated exercise.

The most effective back office e-gov applications are not used simply to facilitate existing workflows but, to reorganize assignments and planning in ways that fundamentally transform government operations integrating work-flow across government in recognition that citizens interact with government as a single enterprise.

Skeptical decision-makers and reluctant public sector managers need to understand and appreciate the value that can be created when technology is used to redesign workflow from an enterprise perspective. While such changes will often be difficult to implement, the potential benefits may very well justify the risks involved. The goal should be to balance risk against return—not merely to minimize risk.

For the aforementioned skeptical decisions-makers, a first step is to recognize and understand the patterns of confusion and conflict that can be associated with a particular e-gov project. The degree in which E-gov can improve a government's administrative systems varies greatly and may be too subjective to measure reliably. The dream of a paperless office is still, for a large majority of government's years away, if it is a realistic outcome at all. The EU report found that only a few of its members confirmed that the expanding use of ICT has had a concrete and tangible effect on administrative reorganization and effectiveness. How new technological tools are used within the framework of existing organizational environments depends on the administrative traditions, practices and cultures endemic to a country. Successful execution of these changes depends on the will which managers and decision makers ultimately demonstrate.

Administrative Issues

E-government program development lacks coordination

There are four fundamental approaches to e-government program development.

1) A nationally coordinated or top-down approach, which is driven by the central government and often features a national strategic plan that coordinates all e-gov initiatives, spending and implementation, among ministries, departments, agencies and units.
2) A nationally autonomous or parallel approach where ministries and agencies develop their own e-gov initiatives with less formal strategic planning, support or coordination from the central government.
3) Sub-nationally or vertically up where local and state governments tend to be the drivers and initiators of programs that rise up and are eventually adopted as policy by the central or federal government.
4) Sub-nationally autonomous approach, where again the innovations and programs are developed at the local levels, but have modest influence on the national governments e-gov activities.

Predominately across ministries, departments, agencies and units e-government development has been autonomous, with only a limited number of countries coordinating national efforts. This could probably be best attributed to the newness of the medium and technology and to the fact that there are few existing policies and strategies to act as guidelines. Although the success of a coordinated e-gov approach cannot be overemphasized, only 35 countries, in 2001, developed a comprehensive official national strategic e-gov agenda and translated it into a government wide policy.

A majority of decision and policy makers choose to proceed from the point of view that e-government should remain part of existing government information technology programs. But developing e-government independent of existing government IT programs has been a key to successful development: the United Kingdom, Ireland and Singapore are examples.

Although coordination from the top down is a strategically sound initiative, in practice actual program development is autonomous and spontaneous in character as most ministries department and agencies go online as their capacity and resources permit. For developing countries, the priority is upgrading internal administrative capacity --- back-office capability --- in order to support e-government programs.

In the developing world, citizen-focused online services barely exists as website content remains static and politicized. Limited resources are an obvious explanation, but a collective lack of confidence and / or creativity on the part of the ICT strategists is another reason. Ultimately the push will come from an informed civil society. The developing countries have been slow to accept a coordinated approach to e-government implementation.

Of the 58 countries in Sub-Sahara Africa, for example, 20 have demonstrated some type of limited coordinated approach to e-government program development. An obvious measure is the websites themselves and how content is displayed.

One important factor impacting all countries, but particularly those where resources are scarce, is the change in thinking required within the public sector, particularly among the administrative culture, when transitioning to web-based service delivery. This actuality is more likely to create a greater sense of unease than most other policies or new programs because of the imposing nature that the technology can project.For example, the decision to assign content-managers the responsibility of keeping information topical and responsive to constituent needs requires an extensive organizational commitment. Also, the delegation of authority that must accompany increased accountability, individual ministries and departments will need to interact more intensively with their clients and the community at large. Traditional processes where material being made public was confined to a specific unit or individual will now need to yield to managerial empowerment, with greater accountability.

The creation of specialized units or divisions mandated with the responsibility of coordinating and implementing the government wide e-gov strategic plan rather than delegating the responsibility to individual agencies remains the exception rather than the rule. E-gov management teams are a significant organizational change and a tangible sign that governments are serious about implementation. They recognize the challenges and realize that the success of projects of this scale depends upon inter-governmental cooperation of an unprecedented scope. These units are usually not an independent agency, fall under the administration of the executive branch and can be ad hoc.

Over 100 countries have official Chief Information Officers, many of Cabinet Rank. Special e-gov offices or task forces appear to be more effective when independent of the CIO. They mandated with the difficult task of launching the e-gov initiatives usually under extremely tight deadlines. Team composition is often an eclectic combination of talent from the policy areas, IT, management and public affairs areas. There are several management models that warrant further study. These include the UK, Ireland, Italy, and to a lesser degree Brasil, Singapore, Australia. The US Firstgov team is operating on very limited resources due to the tragic events of September 11th, 2001.

A prevailing belief which exists among non-IT government staff is that e-gov, especially web-based activities, is merely another policy administrative approach to providing generic public affairs type information. Outside of the ICT community there seems to be a limited sense of e-gov as a major driver of change, administrative reform or reengineering.

Cost effectiveness

The belief that online service delivery is less costly than other channels is not wholly unfounded. But according to a Columbia University Study, this is contingent on several factors like cost per unit / cost per transaction / processing time, etc. Currently, there is very little reliable data to support or refute this assertion. What research or data that does exist is likely to come from local or regional governments. Savings can translate into a source of additional revenue, but most likely this will take several years to realize in both developed and developing countries.

E-services require investments in IT hardware, software and staff. Most e-gov savings expected or anticipated by the transferring and upgrading of service delivery will not be realized in the fiscal year in which the project is launched or the services are upgraded. In fact it will more than likely be several years before an agency or government can show appreciable savings. Planning web-based service delivery programs should include a scheme that automatically collects unit cost data, analyzes and projects costs. According to the Columbia University Study, very little if any microanalysis is being done in the US. Logically, the unit cost of web-bases services will be reduced as there is an increase in citizen use. Accordingly citizen use will increase as more people become web proficient.

The Digital Divide

There is growing concern that e-government will only exacerbate the digital divide and further marginalize the have-nots. Among the reasons for this are, governments prioritize program development based on limited or contracting resources consequently targeting sectors that are more likely to use e-gov. There is a cynical perception on the part of some decision-makers that those without access will never have the motivation or desire to accept and take advantage of e-government programs regardless of what services, hardware or incentives are provided.

Competition among other programs for resources is keen and can result in the more proactive sectors or programs winning out. The question, How long will innovative attempts to bring access to rural poor remain viable and cost manageable is embedded in most planning agendas.

Effective citizen-centric programs prioritize development to reflect a country's ICT capacities, no matter how weak, and user capacity, no matter how limited. If most businesses are connected, the priority may be towards e-commerce service delivery. This is not to suggest that there should be a zero sum attitude toward strategic deployment of e-government programs or that decisions are based on less than equitable factors. It is intended to emphasis the reality that governments face when it comes to allocating resources in choosing the approach that is the most cost-effective and beneficial. Service delivery programs should be built around current technological strengths no matter how limited these strengths may appear to be.

References

Cordella, A 2007. E-government: towards the e-bureaucratic form?, *Journal of Information Technology,* 22, 265–274.

Jeong Chun Hai @Ibrahim. (2007). *Fundamental of Development Administration.* Selangor: Scholar Press.

Koh, C.E., Prybutok, V.R. (2003). "The three-ring model and development of an instrument for measuring dimensions of e-government functions". *Journal of Computer Information Systems* 33 (3): 34–9.

Kushchu I. and Kuscu M. H. 2003. "From e-Government to m-Government: Facing the Inevitable". *The 3rd European Conference on e-Government.* pp. 253–260.

Bibliography

Alavudeen, A.; Venkateshwaran, N. 2010, *Computer Integrated Manufacturing*, PHI Learning.

Allen, T., and M. S. Morton, eds. 1994. *Information Technology and the Corporation of the 1990s*. New York: Oxford University Press.

Amor, D. 1999. *The e-business (r)evolution*. Upper Saddle River: Prentice Hall.

Andrew S. Tanenbaum, 2006. *Computer Networks*, Fourth Edition, Pearson Education.

Attiya, Hagit and Welch, Jennifer (2004), *Distributed Computing: Fundamentals, Simulations, and Advanced Topics*, Wiley-Interscience.

Auslander, Marc A.; Larkin, David C.; Scherr, Allan L. 1981. *The evolution of the MVS Operating System*. IBM J. Research & Development.

Beynon-Davies P. 2004. E-Business. Palgrave, Basingstoke.

Bic, Lubomur F.; Shaw, Alan C. 2003. *Operating Systems*. Pearson: Prentice Hall.

Bynum, Terrell Ward 2008, "Norbert Wiener and the Rise of Information Ethics", in van den Hoven, Jeroen; Weckert, John,*Information Technology and Moral Philosophy*, Cambridge University Press.

Cachin, C.; Guerraoui, R.; Rodrigues, L. S. (2011). *Introduction to Reliable and Secure Distributed Programming*.

Chaudhuri, P. Pal. 2004, *Computer Organization and Design*, PHI Learning.

Chaudhury, Abijit; Jean-Pierre Kuilboer. 2002. *e-Business and e-Commerce Infrastructure*. McGraw-Hill.

Childress, David Hatcher. 2000. *Technology of the Gods: The Incredible Sciences of the Ancients*, Adventures Unlimited Press.

Childs, D. L. 1968, *Description of a set-theoretic data structure*, Technical Report 3 of the CONCOMP (Research in Conversational Use of Computers) Project, University of Michigan, Ann Arbor, Michigan, USA

Codd, E.F. (1970)."A Relational Model of Data for Large Shared Data Banks". In: *Communications of the ACM* 13 (6): 377–387.

Comer, Douglas. 2006. *The Internet book*. Prentice Hall. p. 64.

Cordella, A 2007. E-government: towards the e-bureaucratic form?, *Journal of Information Technology,* 22, 265–274.

Deitel, Harvey M.; Deitel, Paul; Choffnes, David. *Operating Systems*. Pearson/Prentice Hall.

Dyché, Jill. 2000, *Turning Data Into Information With Data Warehousing*, Addison Wesley.

Faber, Jim (1998), *Java Distributed Computing*, O'Reilly: Java Distributed Computing by Jim Faber.

Garg, Vijay K. (2002), *Elements of Distributed Computing*, Wiley-IEEE Press.

Gerstner, L. 2002. *Who says Elephants Can't Dance? Inside IBM's Historic Turnaround.* pg 172.

Hafner, Katie. 1998. *Where Wizards Stay Up Late: The Origins Of The Internet*. Simon & Schuster.

Jesdanun, Anick. 2007. Internet Makeover? Some argue it's time". Seattletimes. nwsource.com. Retrieved 8 August 2011.

Ken North, "Sets, Data Models and Data Independence", *Dr. Dobb's*, 10 March 2010

Koh, C.E., Prybutok, V.R. (2003). "The three-ring model and development of an instrument for measuring dimensions of e-government functions". *Journal of Computer Information Systems* 33 (3): 34–9.

Kurose James F and Keith W. Ross. 2005. *Computer Networking: A Top-Down Approach Featuring the Internet*, Pearson Education .

Kushchu I. and Kuscu M. H. 2003. "From e-Government to m-Government: Facing the Inevitable". *The 3rd European Conference on e-Government*. pp. 253–260.

Manual Castells, 1996. *Rise of the Network Society*, Wiley-Blackwell.

Paul Timmers, 2000. Electronic Commerce - strategies & models for business-to-business trading, pp.31, John Wiley & Sons, Ltd.

Proctor, K. Scott. 2011. *Optimizing and Assessing Information Technology: Improving Business Project Execution*, John Wiley & Sons.

Rainer, R. Kelly, and Casey G. Cegielski. Introduction to Information Systems: Supporting and Transforming Business. Hoboken, NJ: Wiley, 2012. Print.

Reynolds, George. 2009. *Ethics in Information Technology*, Cengage Learning.

Shelly, Gary, Cashman, Thomas, Vermaat, Misty, and Walker, Tim. (1999). *Discovering Computers 2000: Concepts for a Connected World*. Cambridge, Massachusetts: Course Technology.

Silberschatz, Avi; Galvin, Peter; Gagne, Greg 2008. *Operating Systems Concepts*. John Wiley & Sons.

Stallings, William 2008. *Computer Organization & Architecture*. New Delhi: Prentice-Hall of India Private Limited. p. 267.

Tel, Gerard (1994), *Introduction to Distributed Algorithms*, Cambridge University Press.

Webster, Frank, and Robins, Kevin. 1986. *Information Technology—A Luddite Analysis*. Norwood, NJ: Ablex.

William Stallings, 2004. *Computer Networking with Internet Protocols and Technology*, Pearson Education.